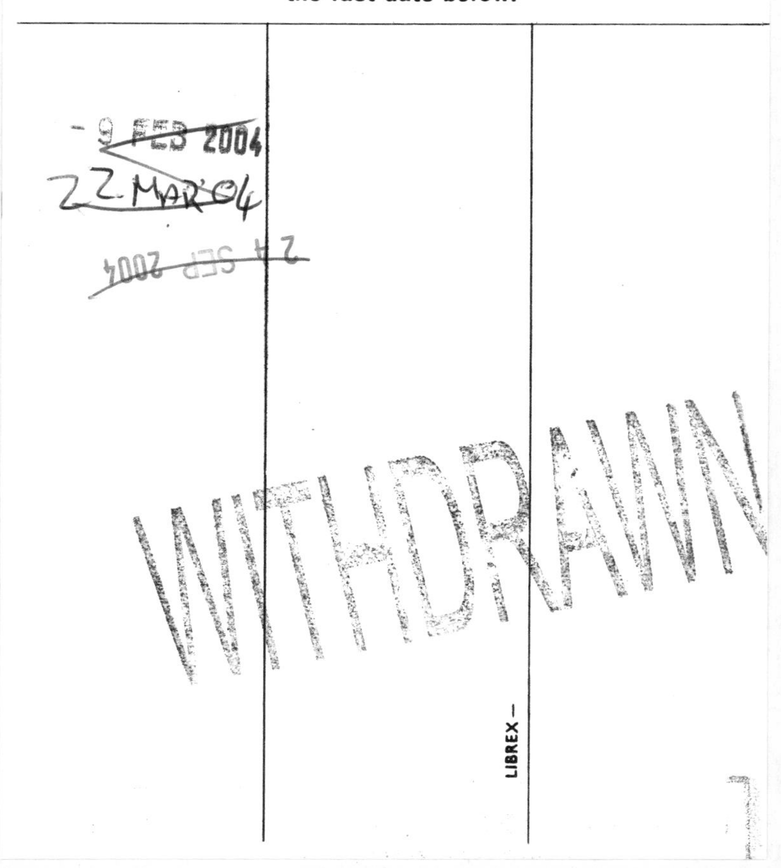
- 9 FEB 2004
22 MAR '04
24 SEP 2004
WITHDRAWN
LIBREX

D1419680

LIVERPOOL JM
3 1111 01026 1442

Law and Social Theory

Series editors:
PETER FITZPATRICK
Professor of Law, Queen Mary and Westfield College
University of London
DR COLIN PERRIN
Faculty of Law, The University of New South Wales

Also available

Law and the Senses
EDITED BY LIONEL BENTLY AND LEO FLYNN

The Wrongs of Tort
JOANNE CONAGHAN AND WADE MANSELL

Delimiting the Law
'Postmodernism' and the Politics of Law
MARGARET DAVIES

Dangerous Supplements
EDITED BY PETER FITZPATRICK

Disabling Laws, Enabling Acts
Disability Rights in Britain and America
CAROLINE GOODING

The Critical Lawyers' Handbook
EDITED BY IAN GRIGG-SPALL AND PADDY IRELAND

The Critical Lawyers' Handbook 2
EDITED BY PADDY IRELAND AND PER LALENG

Family Law Matters
KATHERINE O'DONOVAN

Beyond Control
Medical Power and Abortion Law
SALLY SHELDON

False Images
Law's Construction of the Refugee
PATRICIA TUITT

Contract
A Critical Commentary
JOHN WIGHTMAN

FOUCAULT AND LAW:

Towards a Sociology of Law as Governance

ALAN HUNT and
GARY WICKHAM

London • Chicago, Illinois

To Barnes, Jo, Ros

First published 1994 by Pluto Press
345 Archway Road, London N6 5AA
and 1436 West Randolph,
Chicago, Illinois 60607, USA

Reprinted 1998

British Library Cataloguing in Publication Data
A catalogue record for this book is available from the British Library
ISBN 0 7453 0841 4 hb

Library of Congress Cataloging in Publication Data
A catalogue record for this book is available from the Library of Congress

Designed and Produced for Pluto Press by
Chase Production Services, Chipping Norton, OX7 5QR
Typeset from authors' disks by
Stanford Desktop Publishing Services, Milton Keynes
Printed in Great Britain

Contents

Acknowledgements

We thank the following people for their kind and helpful comments on various aspects of this book: Cora Baldock, Michael Booth, Kerry Carrington, Ian Cook, Patricia Harris, Barry Hindess, Paul Hirst, Russell Hogg, Gavin Kendall, Jeff Malpas, Pat O'Malley, Nikolas Rose, David Silverman, Derek Smith, Bill Taylor, Grahame Thompson and Deborah Tyler. We are extremely grateful to Yolie Masnada for her diligent help with the preparation of the manuscript of Part Three.

Preface

The work of Michel Foucault continues to excite controversy. Passions often run high; his name still produces more partisanship than dispassionate evaluation. It is difficult not to come down either for or against a man who himself delighted in the controversy he provoked. Yet despite the polemics it is already clear, even though his writings have had little chance to gather dust and important parts remain unpublished, that he has become a major intellectual landmark of the twentieth century.

Our book sets itself two limited but nevertheless important objectives. It explores what Foucault has to say about law and it uses his work in the construction of a new framework for the sociology of law. This goes beyond what Foucault himself says and in doing so we attempt to overcome some of the limitations and weaknesses of his work on law.

One of the stimuli for this exploration of Foucault on law is the serious lack of attention his work has attracted from those who make law the central focus of their work. There are two main reasons for this neglect. The first is that law never formed a central interest or focus for Foucault. There is no one book, article or interview that can be directly added to the literature of law. The second reason is more pervasive. Legal scholarship, whether it be academic writing directed at legal academics or texts for law students, exhibits a long-standing intellectual insularity. This is especially true of the Anglo-American tradition of legal scholarship. There is a narrow range of writing that is regarded as relevant. It takes a long time before work generated outside the law schools gains admission even to the portals of the legal academy. Even such a major thinker as Max Weber, who has so much to say that remains of relevance to modern law, is still not routinely part of the literature of law (Hunt 1978). It is little wonder then that Foucault has gained only the most marginal attention within legal scholarship. The analytic bent of Anglo-American legal philosophy barely recognises Foucault as a philosopher, while the empiricist legal history and socio-legal studies pay little attention to Foucault as historian. In so far as Foucault has a presence in the curriculum it is in criminology, but here there is a rather narrow focus on his account of the history of the asylum and the prison.

This book aims to help overcome this absence and to demonstrate the pertinence of Foucault for contemporary issues in legal studies. We stress that the neglect of Foucault cannot be made good by a mere packaging of his direct comments on matters legal. He has a great deal to say about law and even more that has legal relevance, but no amount of stitching together of this material can deliver 'Foucault's theory of law'. Foucault does not have a theory of law. He does *not* have one because law is never one of his major objects of inquiry. Consequently we do not set out to construct a 'Foucaultian theory of law'. But we do contend that an exploration of his work can contribute to the construction of a new and fruitful approach to the exploration of legal phenomena.

This book has three parts. Part One, which is composed of one long chapter, introduces Foucault's thought to a readership which is familiar with law as a significant focus of inquiry but is not familiar with Foucault's writing. It makes no assumptions of prior acquaintance with Foucault's work and offers a brief overview of the significance and implications of Foucault's corpus.

Readers who have a general familiarity with Foucault's work may well wish to proceed directly to Part Two which provides a sustained exposition of the dispersed treatment of law in his texts; this forms Chapter 2. It should be noted that the absence of a systematic treatment of law in Foucault's writings requires us to impose a thematic ordering to the available material. We attempt to remain close to a chronological treatment of the relevant texts but there are occasional departures, for example where he returns in later works to repeat or modify earlier streams. We make no attempt to second-guess how Foucault might have organised a more sustained engagement with law. Part Two does not confine itself to exposition alone. Chapter 3 comprises a critical commentary and evaluation of Foucault's treatment of legal themes. We thereby maintain a clear distinction between exposition and commentary.

In Part Three we present our new framework for the sociology of law. This part of the book is organised around the theme that law can and should be treated as a form of governance. In Chapter 4 we define governance and elaborate this concept by presenting four principles of governance. In Chapter 5 we extend this discussion into the realm of law by making use of our four principles to address law. In this way we outline the content of our new sociology of law as governance. Chapter 6 focuses on method; here we detail the rules we believe the sociology of law as governance should follow. Chapter 7 provides a detailed example of the sociology of law as governance at work; here we outline an approach to the operation of policing.

Part One
Michel Foucault: An Introduction

1
An Introduction to Foucault

Meeting Foucault: some cautionary comments

Foucault's writing on law requires some introduction to the body of his work for those not familiar with his writings. What follows seeks to make Foucault accessible and is largely concerned to give an exposition of those parts of his work which are relevant for the subsequent examination of his remarks on law. We intentionally avoid entering into any extended discussion or criticism of the generality of Foucault's views. We confine our discussion to issues which are pertinent to our subsequent treatment of Foucault on law. Before embarking on an exposition of Foucault's general views it is necessary to note a number of very real difficulties for any introductory overview of his work. There are several reasons why his work does not readily lend itself to easy encapsulation or condensation. We can start by identifying some of these constraints. First, Foucault consciously avoids treating his work as a comprehensive and integrated package. He always insists on the provisional nature of his writings and generally resists the temptation to provide synthesising statements.[1]

His work exhibits a complex self-scrutiny; he persistently revises and reworks a set of themes which emerges and re-emerges throughout his writings, without there ever being clear or unambiguous breaks. This feature is compounded by the fact that Foucault espouses a thoroughgoing suspicion about the possibility of providing guarantees as to the objectivity of knowledge. He does not elaborate a general theoretical perspective or pretend that his concepts are exact or precise. There is no single starting point or grounding of Foucault's thought; it can be approached from a number of different perspectives. One particularly important consequence is that there is no 'real Foucault' who can be summoned. Rather, we argue that it is a useful strategy to insist that there are many 'Foucaults' who coexist and interact with one another.[2] No amount of synthesis can yield a unitary body of knowledge let alone a single theory.

A second difficulty is that it is not easy to place Foucault's thought in regard to other familiar intellectual figures. He deliberately avoids situating his work in relationship to established intellectual landmarks. Both in his historical investigations and in his theoretical interventions he often, but not always, refuses to debate directly with the existing body of scholarship. He generally avoids, for reasons we make clear, the conventional strategy of scholarly writing, namely, to engage in direct debate with those who have previously worked on the field of inquiry before striking out to offer his own interpretation.

It is perhaps even more significant that he avoids making use of concepts used by others. Sometimes this is because he wants to distance himself from the associated body of ideas which have developed around them. This is especially apparent with respect to his relationship to Marxism. It is not that Foucault thinks, for example, that class is an unimportant feature of social structures. Yet, by and large, he avoids analysis in terms of class in order to escape the view that classes are preconstituted social agents with already formed interests and ideologies. This tendency to steer clear of existing intellectual markers and their authors makes the task of situating Foucault in relation to key figures like Marx and Weber fraught with controversy.

Foucault's writings are often 'difficult', particularly on first encounter. It is important to grasp why a certain degree of difficulty and obscurantism is built into his projects. He is consciously concerned to clear the ground of conventional and taken-for-granted questions and approaches in order to explore problems in new and different ways. This rejection of conventional wisdom is apparent in many of his studies. Take the question of the invention of prisons – it has been around as a problem for a long time now. The conventional story with which we are familiar runs something along the following lines: premodern societies used to torture the bodies and cut off the heads of their miscreants until gradually, through a series of imperceptible measures of reform, punishment became less physical, less directed at the body of the offender, and locking up offenders became the most widespread form of punishment. But why? Well, the conventional story runs, it is because people gradually became more civilised and as a result, uneven and complex though the changes were, prisons replaced the gallows. In similar vein we stopped locking up the insane in asylums and began forms of treatment in institutions which exhibited features of both hospitals and prisons. Foucault's work challenges such accounts.

Foucault insists that it is necessary to break with the natural or common-sense reality of the many topics he studies. It is not so much that he denies the existence of insanity or illness, but that he challenges the taken-for-granted self-evidence of madness, sexuality, etc. Thus, in his history of sexuality his starting point is to deny the familiar story that once upon a time sexual life was natural and spontaneous but gradually became subject to various forms of repression whether of guilt-ridden Christianity or of the secular moralism captured by the label 'Victorian'. According to the radical version of this story associated with the counter-culture of the 1960s, the time has now come to liberate ourselves from the repression of sexuality. Foucault rejects what he calls this 'repressive hypothesis'. Foucault himself, it should be noted, sometimes comes close to endorsing radical liberationist views; for example, he comes close to endorsing abolitionist views on prisons and he also comes close to endorsing the view that individuals should be entitled or even encouraged to explore their sexuality to 'the limit'.

A final difficulty surrounding Foucault's work should be mentioned. It does not fit comfortably into any ready-made discipline. His writing straddles the conventional boundaries of philosophy, history and sociology. Frequently commentators with a strong and protective sense of their own disciplinary specialisms resist and strongly denounce Foucault for transgressing their conception of how their discipline should be practised: philosophers suggest that he lacks rigour; historians find him cavalier in his treatment of empirical evidence; sociologists bemoan his lack of precision in defining his concepts and stating his hypotheses. It's not so much that these objections are wrong, but rather that Foucault is self-consciously trying to break down artificial disciplinary boundaries. Such boundaries have their own specific histories; they are not natural or inevitable. Foucault's writings can be read in different ways; he can be read as philosopher, historian or sociologist, but he can also be read charitably, that is with an eye open for what he reveals precisely because he does not allow himself to be bound by the conventional compartments into which knowledge is bracketed. This is how we read him.

Foucault's themes and projects

Now that some of the difficulties in coming to grips with Foucault's writings[3] have been considered, and recalling that we think it sensible to recognise that there are multiple Foucaults, we start by identifying one core theme that runs through his work. Then we introduce a set of major interconnected themes within his writing that are relevant to our subsequent discussion of law. There are numerous commentaries on Foucault that are organised around the progression in his thought and that explore the sequence of the projects on which he writes.[4] A serviceable approach would be to start with his work on madness and the asylum, which takes in his long engagement with the history of the social, psychological and human sciences. Such an account would then move to his work on the institutions of the hospital and the prison and then to the large and uncompleted project on the history of sexuality. The approach we adopt differs. We identify those themes which prepare the way for the engagement with Foucault's discussion of law and related matters. We do not claim that this is preferable to the more closely textual method of treatment. It does, however, allow us to approach our main topic more directly, even though it runs the risk of imposing an unjustifiable unity on the diverse range of Foucault's intellectual production. We take this risk. The best that can be done is to remind the reader of our earlier insistence that Foucault's style involves the deliberate avoidance of any pretence to provide a systematic and integrated body of knowledge or theory.

'Conditions of possibility'

If there is one theme which approaches the status of being central to Foucault's work it is that he is concerned to trace the 'conditions of possibility' of the forms of social knowledge and practices which form the immediate subject matter of his inquiries.[5] As with so much else in Foucault it is as well to approach this topic by inquiring what it is that he is seeking to avoid. The conventional approach of the social sciences is to pursue a causal line of inquiry; to ask what is the cause of the shift or change which stimulates the investigation. Such a causal approach might ask: what caused the shift from the punishment of the body to the imprisonment of the offender? What caused human sexuality first to become the subject of medical intervention and later to come under the sway of the psychological sciences?

Foucault rejects the preoccupation with causes. His rejection is grounded in the tendency of such lines of inquiry to presume that social life is subject to linear and evolutionary change; that the direction of change can be understood as working towards some goal which may at the time have been only dimly perceived by participants but which was nevertheless implicit in the succession of stages that history reveals. The quest for causes tends to introduce assumptions about the role of human intentions, that outcomes are the result of human desires and plans. Foucault thinks that to speak of causes brings with it assumptions that are best avoided.

Foucault's alternative focus on 'conditions of possibility' is self-consciously more modest. It refuses any assumptions about the direction of social change or the role of human plans or intentions. Instead it asks: what combination of circumstances in dispersed and seemingly unconnected fields of social activity combines in such a way as to give rise to some outcome? The kind of inquiry he recommends – genealogical inquiry – manifests a general commitment to the specificity or uniqueness of historical phenomena; it is for this reason that he insists on the 'event'. We will return to this topic below in considering his views on the methods to be pursued in historical inquiries. Foucault's quest for 'conditions of possibility' is closely related to his refusal of the label 'structuralist' which has so often been used to describe his intellectual position. The grounds for his resistance stem from his belief that structuralism involves the idea that structures provide the conditions of their own existence, they provide what is necessary for their own continuation or reproduction. In opposition to this view he is at pains to insist that 'conditions of possibility' are never guaranteed; rather accident and chance play a decisive role. Consider the example provided by modern Darwinism. Writers like Richard Dawkins, in arguing against creationism, concede that the conditions under which life can appear are unique and statistically rare, and that given the enormous time periods of evolutionary history these conditions give rise to the specific combination of circumstances in which unique events occur (Dawkins 1985). In brief, the quest for the conditions of possibility exemplifies a commitment to an insistence upon historical specificity.

Discourse, discursive formation and episteme

Foucault's concern with historical specificity is apparent in his concern to trace the emergence of 'discourses'; how did it come about that some particular way of organising thinking, talking and doing about some selected topic took the form and content that it did?

The use of the term 'discourse' in recent social theory is so pervasive that it is worthwhile discussing this concept and its popularity. In one usage it testifies to the pervasive impact in twentieth-century thought of the 'linguistic turn' which underlies the importance of communication as a distinctive characteristic of social life. At the root of this usage is the basic idea that language does not simply name or label some external autonomous reality except in limited cases such as when I point my finger and say 'chicken'. Most language does more than merely name, such that it is more than a system of direct reference to an external reality. This line of thought does not deny that there is an external reality, but it makes the important point that it is mediated through language. Crucially, Foucault adds institutional practices to this understanding of discourse. The social world is experienced through language and through the ways in which people label and value the context or environment in which lives are lived. Language plays a major part in constituting social subjects, the subjectivities and identities of persons, their relations and the field in which they exist, but only within a context of institutional practices. Thus to be a woman, or a man, or a parent, etc. is not just a biological label, but is encrusted with all the complex things that it means to be a woman, a man, or a parent in a particular culture at some particular point in time. This reading of the concept 'discourse' encourages questions such as: how did it happen that some particular way of thinking, speaking and doing came to prominence at a particular time? Why this way of thinking, speaking and doing rather than another? Discourse emphasises the processes that produce the kinds of people, with characteristic ways of thinking and feeling and doing, that live lives in specific contexts.

Discourse provides a means of designating the different forms of communication, but also of reminding us of the institutional, cultural or constitutive place of language. The term reminds us that words work for us because they are part of some wider phenomenon. While the more important forms of discourse are speech or writing (texts), discourse can also be non-verbal, physical acts (shaking hands) or visual symbols (the genres of film that allow us to distinguish cartoons from westerns). Discourse refers to elements which make up if not always a coherent totality at least a wider frame of reference. Thus a simple social practice such as entering a church where men take off their hats and women keep them on is part of a discourse which we can understand and make sense of involving elements about a sexual division and features of theology; discourse is institutional doing and the language it entails. Discourses put in place a set of linked signs. What the concept captures is that people live and experience within discourse in the sense that discourses impose frameworks which structure what can be experienced or the meaning that experience can encompass, and thereby influence what can be said, thought and done. Each discourse allows certain things to be said, thought and done and impedes or prevents other things from being said, thought and done.

On the basis of this very general introduction to discourse we can now turn our attention to the important directions in which Foucault develops discourse theory. One of his concerns is to ask the question: how do specific forms of knowledge and theory become possible (OoT 1970: xxi)? Much of his attention is directed towards the formation of a series of specialised intellectual discourses. For example, in *The Order of Things* his attention is focused on biology, linguistics and economics, while in *The Birth of the Clinic* (BC 1973) his focus is on the transformations in medical discourse. However, he is not only concerned with such organised or professional discourses. In *Madness and Civilization* (M+C 1965), while paying attention to medical and psychiatric discourses, his emphasis is on those discourses that grapple with the relationship between madness and reason. Again, in *The History of Sexuality* (HoS 1978) he interweaves detailed consideration of theological and medical discourses of sexuality with a broader concern with the role of sexual discourses in everyday life. It is possible to identify a number of features of his treatment of discourse.

Discourses have real effects; they are not just the way that social issues get talked and thought about. They structure the possibility of what gets included and excluded and of what gets done or remains undone. Foucault identifies a distinctive feature of the discourses on sex – that it 'never ceased to hide the thing it was speaking about' (HoS 1978: 53). In its most obvious sense discourse authorises some to speak, some views to be taken seriously, while others are marginalised, derided, excluded and even prohibited. Discourses impose themselves upon social life, indeed they produce what it is possible to think, speak and do.

Foucault is concerned to stress the dispersion and the unruliness of discourses; they proliferate, clash, compete and collide.

> Discourses must be treated as discontinuous practices, which cross each other, are sometimes juxtaposed with one another, but can just as well exclude or be aware of each other. (*Is it Useless to Revolt?* 1981: 67)

It is important to stress that discourse is not simply that which masks or hides; in a sense it is more important than that because it sets up what it is that is argued over and fought about, 'discourse is the power which is to be seized' (*Is it Useless to Revolt?* 1981: 52–3).

Foucault introduces two additional concepts which designate patterns of stability within discourses: 'discursive formation' and 'episteme'. Discursive formation refers to a system of more or less stable elements of discourse that are linked or associated. As a first approximation a scientific theory, a political rhetoric or a theological position are examples. The originality of Foucault's conception is that it involves more than the aggregation of discourses into some relatively persistent field. He provides two advances over the use so far made of the concept of discursive formation. First, he insists that the system of discursive statements which constitute a formation are not merely a unity but also enshrine a 'dispersal' (AoK 1972: 38). Second, the concept focuses attention on its conditions of existence, the conditions that make that formation possible. He shifts attention away from the internal dynamics of the constituent elements employed (signs, signifiers and signified). His account of discursive formations thus breaks with the internal preoccupations of linguistics in order to focus upon the external or social conditions within which discourses are formed and transformed.

The second related concept, 'episteme', has passed into English usage without translation. Episteme refers to historically enduring discursive regularities, the ordering, often unconscious, that gives rise to distinctive forms of thought which underlie the intellectual disciplines. The epistemes provide 'grids' for perception, that is, impose a framework of categories and classifications *within which* thought, communication and action can occur.

Foucault produces some great beginnings. In the *Order of Things* (OoT 1970) he traces genesis of work to a short story by Borges in which he quotes an imaginary Chinese encyclopaedia which classifies animals as follows:

> animals are divided into: (a) belonging to the Emperor, (b) embalmed, (c) tame, (d) sucking pigs, (e) sirens, (f) fabulous, (g) stray dogs, (h) included in the present classification, (i) frenzied, (j) innumerable, (k) drawn with a very fine camel-hair brush, (l) et cetera, (m) having just broken the water pitcher, (n) that look from a long way off like flies. (OoT 1970: xv)

Foucault makes us aware of the centrality of classification and ordering of knowledge by exposing us to a classificatory 'grid' which through the prevailing western episteme is incoherent or even crazy. The epistemes refer to these broad constellations or patterns of thought. He identifies a number of historical shifts or displacements which yield a broad periodisation of western history; successive epistemes are incommensurate and thus mark shifts or ruptures in the organisation of human knowledge. He speaks of two great discontinuities in the episteme of western culture: the first brings about the 'Classical Age' (mid seventeenth century) and the second the 'Modern Age' (beginning of the nineteenth century), 'the threshold of modernity' (OoT 1970: xxiv). We see in Part Two that this periodisation into classical and modern plays an important part in his writing on law.

It should be noted that while Foucault emphasises the importance of discourses he avoids suggesting that the social is coextensive with discourses. He explicitly recognises the existence of a realm of non-discursive elements.

> what I call an apparatus is a much more general case of the *episteme*; or rather, that the *episteme* is a specifically *discursive* apparatus, whereas the apparatus in its general form is both discursive and non-discursive. (P/K 1980: 196–7)

It remains unclear exactly what we are expected to understand the non-discursive elements to be. When he uses the term he refers to 'institution' and to 'apparatus'; the latter he views as an ensemble of discourses, institutions, laws, administrative measures, scientific statements, philanthropic initiatives, etc. He distances himself from his earlier image of the episteme as more or less unitary and coherent frameworks of ideas (OoT 1970: 168).

We suggest that the most fruitful way to understand the relationship between the discursive and the non-discursive is to see Foucault's concern as being with the way in which specific discourses (e.g. medical, legal) get articulated with other social practices external to them. In simple terms, his concern is with the relationship between ideas and practices.

Truth and knowledge

Discourses generate truths, or to be more precise truth-claims. Truth is produced; but it is not produced dispassionately or impartially, it is produced with a passion, with what Foucault calls a 'will to truth' or a 'will to knowledge' and gives rise to regimes of truth.

> Each society has its regime of truth, its 'general politics' of truth: that is, the type of discourse which it accepts and makes function as true; the mechanisms and instances which enable one to distinguish true and false statements, the means by which each is sanctioned; the techniques and procedures accorded value in the acquisition of truth; the status of those who are charged with saying what counts as true. (P/K 1980: 131)

This conception of truth intentionally pits itself against the views of truth that have been prevalent since the Enlightenment whereby truth is neutral, revealing itself only when it is separated from power, in the clear light of day under the scrutiny of scrupulous inquiry. For Foucault, truth is not counterposed to falsity or error, but rather regimes of truth lay down what is true and what is false. Truth operates through the exclusion, marginalisation and even prohibition of other competing truths; indeed it is itself a 'prodigious machinery designed to exclude' (OoD 1981: 55). Truth is not separated from power, rather it is one of the most important vehicles and expressions of power; power is exercised through the production and dissemination of truth.

This view of truth carries with it a risk, the risk of a slide into relativism where everything (and anything) is truth and only power is the arbiter between competing regimes of truth. Foucault denies the charge of relativism. He does so by denying that anything could be true. Here, as in many other places, he shows the influence of Nietzsche, in not pulling back from a profound distrust of all discourses of 'truth' and 'reason'. He certainly holds that the metaphysical quest for truth is doomed, there are no final causes or ultimate truths. Once such quests are abandoned, Foucault recommends that we focus our energies on the modest and realisable task of identifying techniques of truth and power. How is truth produced? What are the conditions of its production? The kind of answer he provides is illustrated by his account of the transformation of medical knowledge and practice; it was the emergence of a regularised and systematic 'clinical medicine' practised within the hospital that laid the foundation for modern medical knowledge. Note that it is not only new medical discourses that are involved, but it is the combination of new discourses, new practices and new institutions which account for the shifts in medical knowledge.

More radical and more disturbing implications arise from Foucault's exploration of the history of sexuality. His target, in *The History of*

Sexuality, is the optimism that gained momentum during the twentieth century which argued that once the guilt and repression which had scarred the discourses on sexuality of the nineteenth century, encapsulated in the term 'Victorian', were displaced, progress towards sexual honesty and liberation could be ensured. Foucault challenges this view which he calls the 'repressive hypothesis'. He does not simply reverse the optimism of the 'sexual revolution' of the 1960s, nor does he join hands with the 'new puritanism' of radical feminism. In important respects his views are more unsettling. The scenario he paints is one of a succession of discourses on sexuality which all aspire to reveal 'the truth of sex', through which we can hope to discover and even master our sexualities and eventually be able to 'tell the truth of sex' (HoS 1978: 57).

Yet there is a disturbing ambiguity in his conclusions. On the one hand he parodies the cults he associates with the 'Californian cult of the self' – through the interrogation of our sexuality we may hope to discover our true selves, to decipher the truth of self and sex through sexual self-inspection and psychotherapy (GE 1984: 362). Yet on the other hand he was himself committed to the exploration of the 'limit-experience' of our sexuality (*Remarks on Marx* 1991: 29).

For present purposes we can suspend the really big questions about the possibility of truth; in the meantime we can draw upon Foucault's impulse towards inquiries which interrogate the conditions of possibility of specific historical truths. What we need to alert ourselves to is that law is one of the more voluble discourses which claims not only to reveal the truth but to authorise and consecrate it. The truth of law is not to be taken for granted but seen as a problem to be investigated: 'it is as if even the word of the law could no longer be authorised, in our society, except by a discourse of truth' (OoD 1981: 55).

Knowledge and power

The linkage between knowledge and power is probably the best-known feature of Foucault's work. The direct linkage, power-knowledge, in a sense says it all.

> We should admit ... that power produces knowledge (and not simply by encouraging it because it serves power or by applying it because it is useful); that power and knowledge directly imply one another; that there is no power relation without the correlative constitution of a field of knowledge, nor any knowledge that does not presuppose and constitute at the same time power relations ... the subject who knows, the objects to be known and the modalities of knowledge must be regarded as so many effects of these fundamental implications of power/knowledge and their historical transformations. (D&P 1977: 27–8)

In order to think through the implications of the concept of 'power-knowledge' it is necessary to put aside what remains the commonplace view, but one which was crucial to Enlightenment thought, that knowledge can only flourish where power is absent, excluded or suspended. Does this mean that we should challenge the idea at the heart of today's human rights discourse, namely, that dictatorships are bad because, among other things, they distort or suppress knowledge? Foucault does not say that the link between power and knowledge is ethically unimportant; rather he insists that it is inescapable. Hence his target is the commonplace assumption in the liberal democracies that all is well whenever and wherever knowledge can flourish independently of power. To understand the formation of any body of knowledge always involves the consideration of the power dimensions within which the knowledge is produced. But this is not to adopt the naive moralism that knowledge is bad, polluted or corrupted by virtue of its production within relations of power and within institutions. Thus he speaks of the prison becoming a 'permanent observatory' and functioning as an 'apparatus of knowledge' directed towards the production of 'docile and useful bodies' (D&P 1977: 136–8); there is an important sense in which the modern prison is more cruel than the old physical cruelties of torture since disciplines impinge on the soul, will or personality of the prisoner.

Once we have appreciated the ideas Foucault invites us to renounce, or at least to suspend, we can then follow his application of the power-knowledge couplet. His point is simple but important: knowledge is a major resource of power. He does not mean that we should simply search out the social interests at play within a power relation, but makes a more complex point that experience always involves some play of power/knowledge; for example, the discourses on human sexuality involve a linkage of power, knowledge and pleasure (HoS 1978: 11). He directs our inquiries towards the 'will to knowledge' that serves as both the support and the instrument of power. The result of this line of thought is to direct attention toward the sites of production of knowledge, the learned disciplines and the professions, in order to unearth their complicity in power relations. Just as discourses exclude or marginalise some other discourses while empowering others, so sites of knowledge also subordinate other knowledge. Politically this leads him to insist that we attend to or listen to these alternative knowledges. For example, the knowledge produced by the modern medical profession has secured the medicalisation of pregnancy as an illness, as a proper place in which to practise medical technologies. In so doing the knowledge associated with the long-standing traditions of midwifery has been marginalised and even criminalised, yet today in many countries, midwives are not merely resisting but asserting their legitimacy. It is, however, interesting to note that the general dominance of the medical model is such that in order to take a recognised place midwifery is having to organise itself as a form of professionalised knowledge and practice.

These concerns give a strongly historical inflection to Foucault's investigations of power-knowledge. These are undertaken in a way which tries to ensure that he is not undertaking a 'history of ideas'. His objection to this way of proceeding is that such an approach always runs the risk of conceiving of 'ideas' as free-floating entities which insert themselves into historical contexts. By contrast, Foucault focuses attention on the way in which the discourses within which knowledge is located are part of the practical tactics and techniques of power relations. For example, in his history of sexuality he identifies four significant strategic shifts since the eighteenth century involving 'specific mechanisms of knowledge and power centering on sex' (HoS 1978: 103). There was first the hysterisation of women's bodies captured in the image of the 'nervous woman'. Then there was a focus on the sexuality of children manifest in the onslaught on masturbation. This was followed by a mobilisation of the 'responsibilities' of married couples to raise families using fiscal tactics to induce an increase in the birth rate. The final stage he identifies with psychiatrisation of perverse pleasures centred on discourses of 'perversions'.

These considerations on the relation between power and knowledge are summed up in the following:

> it is in discourse that power and knowledge are joined together ... we must not imagine a world of discourse divided between the accepted discourse and excluded discourse, or between the dominant discourse and the dominated one; but as a multiplicity of discursive elements that can come into play in various strategies ... discourse can be both an instrument and an effect of power ... Discourse transmits and produces power; it reinforces it, but also undermines and exposes it, renders it fragile and makes it possible to thwart it. (HoS 1978: 101)

Power

Foucault proposes a radically new account of power. Perhaps the best way to explore his account of power is to focus on the view of power that he is trying to escape. He wants to get away from the simple equation of power with repression. He sees this as particularly characteristic of the Marxist tradition. However, one can make his point even more strongly by drawing attention to the pervasiveness of the negative, normative view that 'power is bad'. It is of the greatest importance to stress that he does not try to reverse this commonsense view in order to say 'power is good'. Rather he encourages us to focus on a more analytical and even descriptive approach; we should start by checking out what forms of power are at work in those social situations we seek to understand. It also follows that he rejects a 'zero-sum' view of power by which power on one side always means the other side lacks power or is powerless. Indeed he is concerned to reject the very idea that power is something that is possessed, that some sorts of agents 'hold' power and that others lack it. We should view power as present in all forms of social relations, as something that is 'at work' in every situation; for Foucault power is everywhere. This helps us to avoid one of the least helpful views that stems from the 'power is bad' assumption, namely, that the ideal or goal should be to achieve a situation in which power is absent or abolished. Quite simply, there can never be a power vacuum or a no-power situation or relationship.

It is also important to insist that Foucault is not advocating a view that we should be, or even that it is possible to be, neutral about power. This is very clear from the fact that he continues to stress that the play of power produces systematic power relations, that there are rulers and ruled, dominators and dominated. He is, however, anxious to insist that we should not start out from simple polarities of power (capitalists versus workers, men versus women, etc.). One way of unpacking what he says (although he never puts it in quite this way) is that our analysis should proceed through two stages: first, identify the powers at work and, second, evaluate the results of the play of these powers by making judgements about whether the cumulative effects give rise to domination or subordination. The need to suspend the judgement on power is well illustrated by contemporary controversies over 'body politics'. How should we assess preoccupation with diet and exercise? Do these practices produce increased autonomy and capacity for self-determination? Or are they evidence of an insidious patriarchal power? The political conclusion is controversial; power, in producing the people that we are, is both productive and dominating.

Looked at in terms of this two-stage approach we can see a certain unevenness in Foucault's contribution. He offers fertile and original insights into the mechanics, tactics and techniques of the dispersed forms of power. He is, however, less than clear about what is involved in making the more normative judgements about what constitutes domination and how we distinguish between dominators and the dominated; for example, if women put themselves at risk undergoing cosmetic surgery are they exercising their autonomy or are they the victims of patriarchal power? He simply has little to say on such questions.

This substantive account of power produces a theory of productive power:

> We must cease once and for all to describe the effects of power in negative terms: it 'excludes', it 'represses', it 'censors', it 'abstracts',

> it 'masks', it 'conceals'. In fact power produces; it produces reality; it produces domains of objects and rituals of truth. The individual and the knowledge that may be gained of him belong to this production. (D&P 1977: 194)

One important consequence of his rejection of a negative or repressive view of power is that he insists on the importance of what we may call the 'small powers', what he calls the 'micro-physics of power' (D&P 1977: 12; HoS 1978: 26). These involve the application of detailed techniques for the training of the body by making use of 'micro-penalties', minor punishments such as deprivation of privileges. It is not that he thinks that 'big power' has disappeared. Rather he makes a broad historical generalisation that in modernity 'small power', in particular power located in sites away from the central locations of 'big power' (e.g. the state or capital), has become a defining characteristic of power.

Because he is interested in changing our agenda in discussions of power he has a lot more to say about the 'little powers' than about the 'big powers'. One result is that he highlights the politics of everyday life rather than the big institutional sites of power. For example, Foucault has very little to say about economic power in general and capitalism in particular. He does not talk much about capitalism because to do so would be to take him back on to the terrain occupied by Marxism. His concern to break new ground and to challenge the conventional wisdom of both the left and the right does result in certain rather dramatic silences or absences in his work.

One further feature of Foucault's discussion of power requires mention. While much of the thrust of his criticism of alternative conceptions of power is directed against Marxism he also directs his criticism against liberal conceptions of power. His objection against liberalism is that it overwhelmingly poses the question of power in terms of a distinction between legitimate and illegitimate power. Liberalism has increasingly been preoccupied with seeking to delineate the conditions under which the application of state coercion can be justified. This has been a standard preoccupation of liberal jurisprudence. He makes the perceptive point that this construction of the problem of power, one that he labels the 'juridico-discursive' conception of power, has widespread ramifications. The dichotomous distinction between legitimate and illegitimate finds echoes in other discourses; for example, he notes that the discourses of sex are suffused with preoccupation with the binary distinctions between licit/illicit, permitted/forbidden, etc. Foucault's objection is that to organise discussion of power in this way, through the classification of its results, deflects attention away from the techniques and tactics of power.

Liberalism, like Marxism, is permeated by a negative conception of power.

Foucault's conception of power carries with it the strong insistence that power always involves and engenders 'resistance':

> there are no relations of power without resistances; the latter are the more real and effective because they are formed right at the point where relations of power are exercised. (P/K 1980: 142)

Resistance is not external to power or merely a result of its application (HoS 1978: 95); since power marginalises, silences and excludes, the marginalised, silenced and excluded are always present. One expression of Foucault's political radicalism is contained in his call that we listen to the excluded voices of resistance. As a consequence of his emphasis on local and dispersed power he conceives local resistance as its typical form. In opposition to what he takes to be the Marxist exclusive preoccupation with the grand resistance of revolution, he asserts that

> there is no single locus of great Refusal, no soul of revolt, source of all rebellions, or pure law of the revolutionary. Instead there is a plurality of resistances (HoS 1978: 95-6).

However, it should be noted that his attention to resistance is never as developed or as full as his analysis of power. It remains a field of potential study hardly tapped by Foucault himself. One theme of considerable potential should be noted. One way in which he brings to light the coexistence of power and resistance is with respect to the intimate relation of 'success' and 'failure'. He makes the enormously important point that the success of the prison lies in its failures. The failure of prison to reduce illegalities and actually to serve as a mechanism of their amplification explains the unquenched enthusiasm for experimentation with the detailed forms of incarceration by governments and experts. Yet this supreme failure is paradoxically its success in yielding an ever-present mechanism of 'dividing practices', the shifting categories that first classify the dangerous from the respectable, the mad from the sane, the deserving from the undeserving (D&P 1977: 234–5, 264–5). In Part Three we return to this theme to insist that the interdependence of success/failure is one of the most fertile sources of governance.

The problem of state power

An important question is whether it is possible to retain Foucault's emphasis on the productivity of power but still recognise the state and the condensation of power relations that occurs around it and other institutional complexes, such as banks and multinationals. This problem is more than just a matter of achieving an even-handed treatment of two different species of power. There exists a broad post-Foucaultian consensus that any adequate social or political theory has to take account of both 'big power' and 'little power'. The really difficult issue is to find an adequate way of grasping their mutual articulation and interaction. The weakness of Foucault's project is that in putting 'little power' on to the agenda, he appears to ignore or to understate the importance of the processes that aggregate or condense power in centralised sites. This weakness not only manifests itself in a descriptively inadequate theory but also impedes the generation of adequate political strategy. As Poulantzas expresses it: 'one can deduce from Foucault nothing more than a guerrilla war and scattered acts of harassment of power' (Poulantzas 1978: 149).

It is far from clear whether Foucault himself recognises this problem. It is possible to offer reasonable accounts for the view that he does address this issue and also for the view that he never comes to grips with it. Foucault's negative view of the significance of state power is exemplified in the following formulation:

> The idea that the state must, as the source or point of confluence of power, be invoked to account for all the apparatuses in which power is organised, does not seem to me very fruitful for history. (P/K 1980: 188)

The problem with this rhetoric is that there is, of course, no one who argues that the state accounts for 'all' the manifestations of power. Indeed one of the odd features of Foucault's critique of Marxism is the claim that Marxism exhibits a narrowly state-centred view of power; statism there is in Marx, but this provides no justification for ignoring the much more developed analysis that Marx provides of economic power that is dramatically and revealingly absent from Foucault's own work.

It is clear that Foucault focuses increasingly upon the need to examine how different micro-powers are related, how they become aligned and realigned, and sometimes he refers to such powers becoming integrated into a global strategy of the domination of some class or state.[6] There is no doubt that he recognises the existence of major or global dominations; for example, he suggests that micro-powers

> form a general line of force that traverses the local oppositions and links them together ... *Major dominations* are the *hegemonic* effects that are sustained by all these confrontations. (HoS 1978: 94; emphasis added)

This view makes use of the metaphor of a 'diagram' or a 'vector' of power along the lines of a simple vector model as used in elementary mechanics, in which a number of different forces act in such a way to produce a resultant (or aggregate) force acting in a direction different from any of the originating forces.[7] The great attraction of this model for Foucault is that the forces/powers that act remain autonomous and there is no implication of intentionality or purpose which underlies or explains the direction in which the resultant force/power operates.

In the important essay 'The Subject and Power' he seems to recognise that he has perhaps gone too far in stressing the diffusion of power:

> what makes the domination of a group, a caste, or a class ... a central phenomenon in the history of societies is that they manifest in *a massive and universalizing form, at the level of the whole social body,* the locking together of power relations with relations of strategy and the results proceeding from their interaction. (S&P 1982: 226; emphasis added)

Elsewhere he describes as a 'methodological precaution' the need to

> conduct an *ascending* analysis of power, starting, that is, from its infinitesimal mechanisms ... and then see how these mechanisms of power have been – and continue to be – invested, colonized, utilized, involuted, transformed, displaced, extended, etc., by ever more general mechanisms and by forms of global domination. (TL 1980: 99)

And in perhaps more cautious terms he argues that

> one should not assume a massive and primal condition of domination, a binary structure with 'dominators' on one side and 'dominated' on the other, but rather a multiform production of relations of domination which are *partially susceptible of integration into overall strategies*. (P/K 1980: 142)

However, beyond this point he leaves us stranded. Having most valuably stressed the importance of the dispersed powers and then recognised that these can and do become aggregated into 'overall strategies' and 'global domination' he does offer one concrete example: around 1830 the industrial workers in the northern French town of Mulhouse were subjected to a coherent strategy of domination through dispersed exercises of power in their work, housing, consumption, education, etc. He concedes that this combination of moralising practices effected by a range of different agencies installed the domination of the bourgeoisie, but rejects the idea that the bourgeoisie was a unitary self-conscious subject at work to produce this domination.

In the end what Foucault does is to leave open the question: if there is no unitary class that plans its strategy of domination, how is this result or effect produced? This rather important issue is left unaddressed. It is not that we suggest that there is a readily available solution, but we do suggest it is a question that must be posed. Furthermore it is one that has a bearing on law for it is frequently through the mechanism of legislation that such strategies become visible.

Discipline

The general shift of attention towards 'small power' gives rise to one of the most distinctive of Foucault's preoccupations, that of discipline. His treatment of this key concept has both analytical and historical dimensions. Analytically it identifies the existence of a whole complex of techniques of power that do not rely on force and coercion. Historically it generates his key thesis that discipline becomes the distinctive form of modern power. He focuses on the rise in the eighteenth century of new ways of controlling and training people, what he calls 'technologies of the body'. He gives the example of military training the aim of which was to produce 'docile bodies' by means of a 'new micro-physics of power' through the repetition of detailed tasks epitomised by marching drill (D&P 1977: 139). Perhaps the most distinctive embodiment of 'discipline' to which Foucault draws attention is the technique of 'surveillance'; more generally he draws attention to the methods of observation, recording and training. The practices that had long been enshrined in monasteries, armies and workshops became generalised and came to permeate everyday social life.

The disciplines are characterised by these tiny, everyday, physical mechanisms, by systems of micro-power.

> The chief function of disciplinary power is to 'train' ... Discipline 'makes' individuals; it is the specific technique of a power that regards individuals both as objects and as instruments of its exercise ... It is not a triumphant power ... it is a modest, suspicious power. (D&P 1977: 170)
>
> The formation of the insidious leniencies, unavowable petty cruelties, small acts of cunning, calculated methods, techniques, 'sciences' that permit the fabrication of the disciplined individual. (D&P 1977: 308)

These contrast sharply with the 'majestic rituals of sovereignty or the great apparatuses of the state'.

Disciplinary power exhibits three general characteristics. The first of these is hierarchical observation (sergeant over recruits, teachers over pupils). Foucault extrapolates from his work on the prison in which 'surveillance' provides him with the key to the disciplinary regime of incarceration. We may note the link between surveillance in *Discipline and Punish* and 'the gaze' in *The Birth of the Clinic*. Discipline requires detailed observation and the individuation of its 'targets' (prisoners, patients, pupils, etc.). The cumulative observation of large numbers of targets provided an impulse towards the keeping of records, the writing of reports, and monitoring and inspection, all of which came to form important techniques of government in the modern world.

Second, discipline operates through 'norms', normalising judgements (norms are specified which define the attributes of 'good soldiers' or 'obedient children'). The norms are directed against a wide range of behaviour involving faults, such as lateness, untidiness, uncleanliness, disobedience, but they are also directed against faults in attitude such as insolence, disobedience, intransigence, lack of loyalty or team spirit. Norms specify the goals to which those subjected to discipline must strive to achieve – standards of tidiness, punctuality, etc. Foucault refers to these faults as 'offences' but this term rather misses the point he wants to make – in English at least. It is too close to the terms of criminal law. 'Fault' is to be preferred because it brings out the role of broad normative standards. Characteristic of disciplinary techniques are 'exercises' and other forms of the repetition of tasks. It is training which lies at the heart of the disciplines; this feature is perhaps best exemplified in the techniques of nineteenth-century schooling with the emphasis on the repetition of routinised tasks epitomised in the learning of multiplication tables by rote. Cumulatively training uses the 'examination' or other form of test as one of its evaluative techniques.

Third, discipline deploys not so much punishment but a mix of micro-penalties and rewards. 'At the heart of all disciplinary systems functions a small penal mechanism' (D&P 1977: 177), an 'infra' or 'micro-penality' that takes possession of ever-widening fields of behaviour. The difference between disciplinary faults and criminal offences is marked by the typical forms of punishment. Characteristically, the repetition of training assignments forms the core of punishment ('Write out your tables again'). The micro-penalties also deploy forms of minor individualising humiliation ('Stand in the corner', 'Wear the dunce's cap', etc.). They escalate through graduated stages from loss of minor privileges ('No play until you've finished your lessons') to sanctions that mimic criminal penalties (fines for lateness in the factory, and 'minor' beatings in schools). The disciplinary sanctions are significantly linked with the use of 'rewards' (stars, grades, prizes, badges, privileges, etc.).

The advance of disciplinary techniques is manifest in the rise of 'regulation' as a distinctive technique of government. It should be noted that although Foucault occasionally uses the term 'regulation', it is not treated in any systematic way such as to become a concept of any importance in his work.[8] However, the term 'regulation' is important because it sets up a contrast with 'law'; if law is the stipulation of general rules then regulation is more task oriented and less prohibitive, in that it is employed to define detailed goals and targets for training and other forms of intervention directed at the behaviour of individuals. Regulation characteristically involves techniques of detail. In the light of these considerations we make more use of the term 'regulation' than does Foucault himself; we think this usage is entirely consistent with the spirit of his work.

The problem of disciplinary society

Some care is necessary with regard to the historical thesis which is undoubtedly present in Foucault, namely, that the accumulative effects of the diffusion of disciplinary mechanisms led to the emergence of a distinctive 'disciplinary society'. His contention is contained in his claim that the disciplines supplant and even replace law as a primary mode of government. Such a thesis clearly has important implications for our study of law. We return for a more detailed discussion of these issues in Part Two. For the time being it is sufficient to note that the contention that discipline supplants law lies at the heart of our criticism that Foucault tends to expel law from any major role in modern forms of government. What separates Foucault's position from ours is that he counterposes law to regulation, while we see discipline and law supplementing each other and forming distinctive and pervasive forms of regulation at the very heart of modern government.

However, Foucault is somewhat ambivalent in his formulations. Sometimes he asserts a strong claim about the emergence of a new state of 'disciplinary society' and at other times he backs off from this view. In describing the advent of the modern world he suggests that from the eighteenth century the mechanisms of sovereignty were supplemented by 'the invention of a new mechanism of power'.

> This new type of power, which can no longer be formulated in terms of sovereignty, is, I believe, one of the great inventions of bourgeois society ... This non-sovereign power, which lies outside the form of sovereignty, is disciplinary power. (TL 1980: 105)

But at much the same time he explicitly denies that there was a shift from the classical society (organised around sovereignty) to a disciplinary society.

> We must ... see things not in terms of the substitution for a society of sovereignty by a disciplinary society and the subsequent replacement of a disciplinary society by a governmental one; in reality we have a triangle; sovereignty-discipline-government, which has as its primary target the population and its essential mechanism apparatuses of security. (G 1979: 18–19)[9]

There is no easy way to resolve whether or not Foucault holds that modernity is a disciplinary society. It is important to bear in mind that he is attracted to broad developmental theses and yet, at the same time, feels compelled to deny them because of his rejection of totalising theoretical generalisations. Internal tension can be resolved by suggesting that Foucault holds that 'disciplinary society' is a formula which draws attention to the undeniable expansion of disciplinary power, but he does not want to suggest that this resulted in the formation of some systematically disciplined and ordered society.[10]

From discipline to self-discipline: from power to ethics

One of the most important implications of Foucault's concept of discipline is that it paves the way for a significant extension of his capacity to engage with the complex ways in which power is inscribed in social life. Linguistically it is a small step from 'discipline' to 'self-discipline'. Two initial implications of this move should be noted: first, that it is consistent with all his efforts to break with the identification of power with repression and coercion; second, it marks another stage in his break with the Marxist concept of ideology, or at least those versions which emphasise ideology as deception or 'false-consciousness'.

One immediate qualification needs to be entered. The shift that Foucault actually makes does not involve the term 'self-discipline'. He never settles comfortably with any one concept, but rather makes use of a variety of terms: conduct of the self, practices of the self, self-control, techniques of the self, technologies of the self, techniques of self-mastery among others. All these terms are used as he comes to focus his attention, in the second and third volumes of his history of sexuality (HoS 1985a, b), on how the self constituted itself as a subject, 'the history of how an individual acts upon himself' (1988a: 19). He is interested in the ethical problem central for men (and not women, children and slaves) in Classical Greece of 'enkrateia', the project of self-mastery, for example, to be able to look at a beautiful girl or boy without desire. Whereas today art refers overwhelmingly to objects, for the Greeks the issue was to make one's life a work of art.

Self-discipline and the complex of practices, techniques and technologies associated with it should not be thought of as a purely historical set of questions. As always with Foucault the past and 'history of the present' are intimately linked. Today we place enormous investment in the techniques of the self, whether concerned with body-weight, personal appearance, fitness or all the other self-constructing or self-constituting activities. These practices should not be thought of as simply self-directed or introspective; they bring into play a whole series of 'specialists'. Not only are there an array of specialisms based on the 'psych sciences' (Rose 1989), but as a recent *Yellow Pages* advertisement tells us, everyone needs to have their own aerobics instructor, their hairdresser, and their 'tae kwon do' coach.

Foucault's attention to the processes whereby people make themselves involves more than a move from discipline to self-discipline. It provides him with a way in which he can address the important ethical issues involved in his own life. It also gives him a way of securing some linkages between his historical and sociological concerns on the one hand, and his ongoing engagement with those strands in philosophy that require us to decide how we should make ourselves, on the other. In some of his last writings and interviews he brings the strands of his work together with the suggestion that what he has been on about all the time is the interconnections between three primary issues: power, truth and ethics. In different phases of his writing one or the other of these elements comes to the fore, for example the period of *Discipline and Punish*, the mid 1970s, is the period of 'power'. For present purposes it should be emphasised that when we consider each and every form of self-government all three elements are involved: first, the dimension of truth through which we constitute ourselves as subjects of knowledge; second, the field of power through which we constitute ourselves as subjects acting on others; and third, ethics through which we constitute ourselves as moral agents (GE 1984: 351). Thus, relations of power exist alongside the production of truth and, in turn, are linked to questions about how we are to live.

Government and governmentality

The concepts 'government' and 'governmentality' are important and original features of Foucault's thought. As he explains:

> This word [government] must be allowed the very broad meaning which it had in the sixteenth century. 'Government' did not refer only to political structures or the management of states; rather it designates the way in which the conduct of individuals or states might be directed: the government of children, of souls, of communities, of families, of the sick. It did not cover only the legitimately constituted forms of political or economic subjection, but also modes of action, more or less considered, which were designed to act upon the possibilities of action of other people. To govern, in this sense, is to structure the possible field of action of others. (S&P 1982: 221)

The focus on government as an intensely practical matter, 'how things get done', is of great importance and it is this strand, more than any other in Foucault's work, that this book sets out to develop and extend. However, it should be noted that one important implication of Foucault's conception of government is that it is consistent with his downgrading of the importance of the state.

> Maybe, after all, the State is no more than a composite reality and a mythical abstraction whose importance is a lot more limited than many of us think. Maybe what is really important for our modern times ... is not so much the State-domination of society, but the 'governmentalisation' of the State. (G 1979: 20)

His refusal to accord great significance to the state is, in large part, a reaction against the Marxist tradition which he believes places too much significance on the state.

He puts forward the historical thesis that during the eighteenth and nineteenth centuries the practices of 'government', conceived in this expanded sense, came to the fore; the traditional practices of state sovereignty did not disappear, rather the new forms of governmental rationality became more important. One of his most important illustrations is the way in which 'population' came to constitute a central focus for a variety of projects of government; population not only became the target of formal governments, but also of a variety of other governing agencies. Medicine, religion, education and other mechanisms became concerned with the number, health, education and the productivity of the aggregated individuals and organisations that made up a population.

This 'non-governmental' government of population is well illustrated in Valerie Fildes's study of wet-nursing, the employment of women other than the natural mother to breast-feed babies. In England in the mid eighteenth century the trustees of the London Foundling Hospital became concerned by the mortality rate of the increasing number of abandoned children in their care. Not only did they collect detailed statistics to compare the mortality rates of 'wet' versus 'dry' nursed babies (using milk substitutes), but they found that wet-nursing, today regarded as primitive, was in fact more efficient than dry-nursing. What indicates an instance of the government of population is that the hospital proceeded to lay down detailed regulations for the conduct of the wet-nurses it employed; these rules focused in particular on their sexual practices and consumption of alcohol, but also revealed a sense of the importance of diet (Fildes 1988: 160–88). Only much later did the state itself intervene through legislation. Such is 'government' in the Foucaultian sense.

There is no doubt about the historical importance of the shift to the focus on the government of population. Two additional points should be noted. The first is that it is clear that a focus on population had emerged significantly earlier than Foucault suggests. Populationist politics are to be found playing a key role, for example, in the Renaissance cities of northern and central Italy. The city government of Florence made frequent efforts to regulate the size of dowries since high dowries led men to postpone marriage and this in turn led to declining birth rates. The second point is that other strategic governmental targets played a key role in modernisation alongside the concern with population. One long-lasting governmental focus was on 'improvement', which first emerged in Reformation Europe involving not only issues of population but also wider issues of economic development, education and religion. The focus on improvement persistently reappeared down into the nineteenth century. Similarly the preoccupation with 'civilisation' played a decisive role, especially in the era of colonialism and later in connection with the government of indigenous people in North America, Australia and elsewhere. These reservations do not undermine his most general point about the emergence of distinctively new practices of government, but they do open up a space for a more extended debate about their timing and sequence (a point we return to in Part Three).

The thrust of Foucault's work on government can perhaps best be brought out by focusing for a moment on his concept 'governmentality'. He seeks to draw attention to the emergence of new and distinctive mentalities of government or 'governmental rationality' which involved a calculating preoccupation with activities directed at shaping, channelling and guiding the conduct of others (Gordon 1991). Sometimes such activities are undertaken by traditional organs of government. Such activities are well illustrated by, but are by no means limited to, the distinctive nineteenth-century practices of the appointment of inspectors. Karl Marx and many others after him have drawn attention to the enormous importance of the factory inspectors in England. In a more recent study Bruce Curtis paints a vivid picture of the role of school inspectors in Canada West (eastern Ontario) in undertaking projects directed at the 'improvement' of the population during the mid nineteenth century (Curtis 1992). Foucault himself frequently uses the example of the medical profession as an important source of 'government' activity that becomes particularly apparent in periods of significant shifts in the organisation of medical practice, as traced in *The Birth of the Clinic*, or in response to plagues and epidemics. The expansion of medical practice often went hand in hand with concerns about the size and health of populations, and accelerated the practices of 'political arithmetic' (censuses, demography, etc.) and 'public health' (provision of water supplies, sewage systems, etc.); this combination of increasingly organised and coherent medical interventions is well captured by the phrase 'medical police' (Rosen 1974).

One common feature of these examples of informal-governmental government is that they involve, directly or otherwise, the production, dissemination and utilisation of knowledge. Recording, counting, tabulating, calculating, comparing have become both the means by which governmental intervention expands and one of its chief by-products. In the first instance much of the production of such empirical and statistical knowledge was carried out by private individuals and learned societies; only later did these become part of the data-gathering state agencies. The importance of such practices is captured in the phrase 'rule-by-records and rule-by-reports' (Smith 1985). This discussion of the connection between government and the production of knowledge serves to underline the general importance of Foucault's insistence on the integration of knowledge and power, hence power-knowledge.

Not only is government an activity of institutional and quasi-state bodies, it takes place in everyday practices, as is well illustrated by the case of child-rearing practices which, while highly individualised, exhibit distinct patterns that are heavily influenced by a succession of child-rearing experts.

There is little doubt that Foucault's concept of governmentality is extremely suggestive. This should not however blind us to certain obvious weaknesses in Foucault's use of it. As we hinted above, he is very undecided about the identification of specific periods of the historical development of government. He does offer periodisations, but he also drops and changes them. More significant perhaps is that he has surprisingly little to say about the content of the 'mentalities' of modern government. In this respect Foucault contrasts rather poorly with Max Weber who provides a very carefully drawn picture of modern bureaucratic rationality. Foucault should be regarded as pointing us in a direction of inquiry rather than having completed that inquiry. What kind of governmentality is associated with the passion for inspection which is so characteristic of mid nineteenth-century government? How did the self-consciously discretionary government of 'cases' come to the fore in the mid twentieth century?

Closely related to the conception of government is another 'expanded' concept, namely, police. While today we tend to think of police as an institution directed towards law enforcement, Foucault revives a much older and broader conception of 'police' whose object is to 'foster citizens' lives *and* the state's strength' (*Is it Useless to Revolt?* 1981: 252). This view of policing involves a much wider range of social interventions directed towards ensuring a productive and effective citizenry which embellishes the stature of the state. Thus the terms 'police' and 'government' are closely associated; we return to this topic in Part Three.

Strategies, programmes, policies and tactics

Foucault's concern with the practices of diverse agencies that go to make up governmental activity creates a particular problem concerning interpretations of his work for those, like ourselves, not raised on continental philosophy. It concerns the part played by intentional or purposeful activity in history. His resistance to the methods of orthodox historical scholarship lead to his reluctance to accept that outcomes can be understood as the result of intentional, preconceived purposes. Since he is anxious to emphasise the part played by diverse but uncoordinated agencies, he is suspicious of any accounts which imply the existence of some directing hidden hand or of some coordinating agent, whether it be 'the state' or a 'ruling class' which, as it were, masterminds the disparate interventions of many different agencies. On the other hand, he does not think that the trajectory of social change is simply a matter of accident, since he claims to identify distinct 'stages', for example, of forms of government or of medical practice. The upshot is that the conventional terms available to us for linking the purposes and intentions of agents, such as 'strategies', 'programmes' and 'tactics', to the results or outcomes of historical change of social action either have to be rejected or, if they are to be used, have to be used in a very different way.

This problem of interpretation is highlighted by another feature of his general intellectual perspective. He shares with many of his contemporaries a hostility to 'humanism'. Some care needs to be taken in order not to misunderstand what is involved. To be 'anti-humanist' does not mean to be opposed to humanitarian causes and values or to deny the importance of 'human rights'. What is involved is a general methodological concern. Social thought since the Enlightenment has been dominated by 'humanism' in the sense that it takes the individual, conceived as a well-willing and intentional actor, to be the primary agent of social action. Foucault shifts the emphasis by refusing to see the individual, always referred to in these discussions as the 'subject', as the point of origin; rather the subject is the result or outcome of social life. 'My objective ... has been to create a history of the different modes by which, in our culture, human beings are made subjects' (S&P 1982: 208). The subject, with all the attributes of self-consciousness, is not the cause or author of history, but rather its result.

> One has to dispense with the constituent subject, to get rid of the subject itself, that's to say, to arrive at an analysis which can account for the constitution of the subject within a historical framework. (P/K 1980: 117)

The subject is always produced, is the result of social processes which give rise to the 'subjectivity' within which individuals experience themselves. Foucault takes the notion of the subject through two further stages. First, he is at pains to insist that subjects are not simply the result of external social forces, but rather are actively engaged in their own production; we produce ourselves as 'selves' or as identities. His second extension uses a significant feature of the word 'subject'; it has two senses, people are both subjects (self-conscious beings), but they are also 'subjected' (power acts to produce 'subjection'). As he puts it: 'An immense labor to which the West has submitted generations in order to produce ... men's subjection: their constitution as subjects in both senses of the word' (HoS 1978: 60).

Foucault's resistance to all analyses which start with the subject has resulted in many commentators describing his position as 'structuralist'. Structuralism is the view that there are objective processes which exist independently of consciousness and which produce subjects. Foucault strenuously denies the label structuralist since he actively rejects the existence of processes that lie outside consciousness. It is for this reason that we describe his position as anti-humanist, although it should be borne in mind that he does not himself use this term.

We now return to the question of how Foucault deals with the issue of the role of intentionality or purpose in human action. He states what at first sight is a paradox:

> Power relations are both intentional and nonsubjective ... they are imbued, through and through with calculation: there is no power that is exercised without a series of aims and objectives. But this does not mean that it results from the choice or decision of an individual subject. (HoS 1978: 94–5)

This point is of great importance because unless we keep it in mind it is all too easy to read Foucault as saying that the modern forms of disciplinary power make up a monolithic regime of all-embracing and all-consuming power. While he does not always say it as clearly as he might, he holds that the three elements of power, its discourses, its practices and its effects, never fit together or correspond (Gordon 1980: 246–55). This point can be seen more concretely in the context of his discussion of the success/failure of the prison; while the discourses of incarceration speak of reforming the inmates, the practices and effects of prison life 'produce' offenders. In brief, Foucault argues that while prisons are advantageous for capitalist interests, this was 'discovered' rather than intended or planned.

Power for Foucault never produces the effects its discourses promise. This point has a similarity with an important sociological issue about the gap between 'intentions' and 'unintended consequences' of social action, found, for example, in the work of Robert Merton and C.W. Mills. Much orthodox sociology has used this insight to set up a practical project of the social sciences to close the gap between policy and outcome. Foucault strives to go beyond this pragmatic concern with 'unintended consequences' by insisting that the very nature of power, its success/failure, lies in what we might call the necessary non-correspondence between discourse, practice and effects.

One of his most interesting but tantalising approaches to this question is his discussion of 'strategy'. He uses the term in a very distinctive way. He proposes that in specific historical conjunctures combinations of plans or programmes, distinct forms of knowledge and particular practices 'come together' so that it is possible to identify the existence of a strategy, for example the strategy of rehabilitation in mid twentieth-century penology. At the same time, he refuses the idea that strategies are coherent vehicles of the intentions of some identifiable social agent such as a class or a party. He is, for example, very suspicious of any talk about the 'strategy' of the capitalist class or even of governments. He attempts to drive his point home, in a way which in the long run probably confuses rather than clarifies, by suggesting that strategies can exist without there being any 'strategists'.

There is no doubt that the issue he addresses is important and complex. It is one that has a particular significance for our concern with law for two significant reasons. First, law seems to provide strong evidence of the existence of strategies. For example, legislation such as the reform of the poor law in Britain in 1834 or case law such as *Brown* v. *Board of Education* (1954) in the United States concerning civil rights seem to provide evidence of moments of significant strategic shifts.[11] Second, such legal change suggests that at particular historical moments law reflects or incorporates an aggregation or condensation of shifts in the disposition or direction of power.

Since this issue of 'strategy without strategists' has special relevance for our understanding of law we have highlighted this issue and we will return to it again in Part Two. For present purposes it is sufficient to note that we think it is important to pose the question of whether the state and the legal system are significant institutional locations at which power becomes aggregated or condensed. Sometimes Foucault seems to make precisely this point. In a very characteristic formulation he argues that

> domination is organized into a more-or-less coherent and unitary strategic form; that dispersed, heteromorphous, localized procedures of power are adapted, re-enforced and transformed by these *global strategies* ... a multiform production of relations of domination that are partially susceptible of integration into *overall strategies*. (P/K 1980: 142; emphasis added)

But more often his preoccupation with the heterogeneity of power and its tactics, its dispersion and its capillary nature leads him to ignore this aggregation and even to deny it. While everyday power is undoubtedly crucial it is also important to keep under consideration the fact that the diffuse techniques of power sometimes come to be aggregated in the massive institutional presence of state, legal, military and economic apparatuses. The limitations of Foucault's treatment of 'strategy' stem from his insistence on the diversity of power relations while at the same time rejecting both structural determination and the existence of objective interests, such as those of classes or institutional apparatuses. The result is that he is left with no means of accounting for the aggregation or globalisation of power. To talk of strategy as he does is to imply some principle for the historical patterns of power relations without providing the means to offer an explanation of their specific manifestations.

Let us consider his analysis of the strategies at play in the regulation of sex and sexuality. He argues that in the nineteenth century the discourses of sexuality were dominated by medical concepts and analogies but that these were replaced by psychiatric concepts and analyses in the twentieth century. The challenge is to explain why it was that this movement occurred, how it went in one direction rather than another. Despite his invoking of 'strategy' his general theoretical stance impedes him from providing any but the most gestural account of how this occurred. Boaventura de Sousa Santos criticises Foucault along similar lines when he suggests that Foucault simply goes too far in stressing the dispersion and fragmentation of power and that this results in a lack of attention to the way in which hierarchical patterns in the forms of power emerge and then change to different configurations (Santos 1985).

One alternative to Foucault's treatment of strategy is provided by Poulantzas who seeks to avoid the idea of a unitary, omniscient, omnipresent 'Power-State'. While he suggests that the state is indeed potent and ubiquitous, at the same time he draws attention to the prodigious incoherence and chaotic character of state policies. 'Strategy', in this view, emerges only after the event through the collision of different tactics in which the general pattern of change is the complex result of the balance of forces produced when the specific tactics of a variety of social movements and classes clash and compete (Poulantzas 1985: 135–7).[12] It is worth stating that though we suggest the limitation of Foucault's neglect of the condensation of power, particularly of state power, is significant, this does not detract from the exciting potential opened up by his discussion of power and strategy.

History, archaeology and genealogy

Foucault exhibits a deep caution about the nature of the historical project such that he denies that what he is doing is 'history'. He is reacting against certain assumptions that have come to inform the orthodox academic discipline of history. The problem with history revolves around the way in which the past and the present are conceived. The simplest illustration is the tendency for historical scholarship to read the historical processes as the unfolding of the present in the past; in this way conventional history focuses on those features of the past which contribute to those elements which come to form the present while downplaying those aspects judged not to be significant to the present. For example, the history of political institutions is written by focusing on those features which play a role in modern institutional forms of parliamentary democracy while the elements that do not fit this model are treated as archaic, or ignored. In broad terms it is the persistent evolutionary assumptions about the connection between past and present which offend Foucault. As we have already seen, his general strategy is to insist upon the uniqueness or specificity of 'events'.

Foucault's espousal of first 'archaeology' and later 'genealogy' is a self-conscious attempt to escape from the pervasive features of orthodox history, such as the assumptions of linearity, teleology, evolution. We concentrate here on his method of genealogy and leave aside the changes in his position which led him to move on from his earlier concern with archaeology. For our purposes, both archaeology and genealogy share the same general principles. Genealogy accords priority to events by focusing on their singularity or particularity (and thus on their heterogeneity, taking account of the role of accidents, errors, etc.).

His most general definition of the genealogical method identifies it in the following terms:

> What it [genealogy] really does is to entertain the claims to attention of local, discontinuous, disqualified, illegitimate knowledges against the claims of a unitary body of theory which would filter, hierarchise and order them in the name of some true knowledge ... Genealogies are therefore not positivistic returns to a more careful or exact form of science. They are precisely anti-sciences. (TL 1980: 83)

The significant claim that he makes for his method is that it allows access to those features which get pushed aside and forgotten in orthodox history.

> It is through revolt that subjectivity ... introduces itself into history and gives it the breath of life. [He gives the examples of resistance by delinquents and the insane] ... One does not have to maintain that these confused voices sound better than the others and express the ultimate truth ... it is sufficient that they exist and they have against them so much which is set up to silence them ... it is due

to such voices that the time of men does not have the form of an evolution, but precisely that of a history. (*Is it Useless to Revolt?* 1981: 8)

This passage shows his effort to escape from a linear view of history conceived as sequences of cause and effect. Rather, he insists that if history is to show how we have become what we are (hence his description 'history of the present'), then it is about the 'small' happenings, not themselves part of any master plan or subject to any grand design.

This approach to historical events ties in with his vectoral analysis of power described above. Disparate forces act upon social objects and the end result (the direction in which the object moves) is the outcome of that totality of disparate forces, but is a direction distinct from any of the constitutive forces at work. Still with this analogy, if we regard the vectors as the intentions of social agents, the resultant vector is not the coordinated result of any particular intention(s), but is, necessarily, radically contingent. Outcomes are limited by the 'field of possibility' within which the action is situated. In brief, the direction of history is contingent.

Foucault and Marxism

A brief discussion of Foucault's relationship to Marxism is necessary because many features of his intellectual trajectory are reactions to the significant influence of Marxism on French intellectual life. After a brief period of membership of the French Communist Party (PCF), he made an early breach with 'the Party' and exhibited increasing hostility toward the Stalinist politics that he saw it representing. Theoretically he treats Marxism as a tradition irretrievably marked by economic reductionism. The paradoxical result was that in the 1960s and 1970s when western Marxism was undergoing its most fertile development and breaking, among other things, with a narrow equation of power, state and repression, we find Foucault invoking what we might call 'Marxism at its worst' as the intellectual spur driving forward his reconceptualisation of power.

A central feature of Foucault's project lies in the distinctive form of his engagement with the legacy of Marx.[13] As is so often the case in the history of debates with the 'ghost of Marx', they occur tangentially and in fragments. This is certainly true in Foucault's case. While he insists on the impossibility of undertaking historical scholarship without taking serious account of Marx, most of his other scattered comments are more negative (P/K 1980: 52–3). What Foucault does is to 'use' Marx to set up a negative pole against which to elaborate his alternative. The Marx that emerges is somewhat one-dimensional:

rigid determinist, economistic, with a narrow conception of power as repression, and viewing the state as a unitary agent and the instrumental bearer of the interests of the ruling class.

The other, and more interesting, facet of Foucault's treatment of Marx is the conscious avoidance of Marx's concerns and his concepts. This he does not because he rejects these concerns but rather in order to avoid being trapped within the terms in which they have been debated in the Marxist lineage. Foucault's relationship with Marx can best be understood as a self-conscious avoidance of Marx, but it is an avoidance that should not be mistaken as an ignorance of Marx. By 'avoiding' Marx we should understand Foucault to be opening up new and unencumbered ways of addressing both new and classical problems. For example, Foucault self-consciously avoids dealing with the question of the state. 'I don't want to say that the State isn't important; what I want to say is that relations of power ... necessarily extend beyond the limits of the state' (P/K 1980: 122). Similarly, he avoids Marx's concept of ideology for fear that it leads to the espousal of an opposition between truth and falsity; thus Foucault's variant of the concept of discourse fills some part of the space vacated by the concept ideology. One fruitful way of understanding Foucault's work is to recognise his strategic reasons for *avoiding Marx*.

However, there is a price to be paid for his concern to circumvent a negative conception of power. His critical step is the equation and conflation of negativity with repression; the result is that in order to avoid a negative conception of power he first downplays (but does not exclude) the repressive capacity of power and then proceeds to elaborate an account of the modern forms of disciplinary power which is founded on non-repressive forms of domination. In order to secure this objective he sets out to purge all those elements associated with negativity and repression. This has two significant consequences. In the first instance, as we have seen already, he displaces the question of the state because he views the Marxist problematic of the state as inescapably bound up with the equation of class power and repression. Second, as we examine in more detail in Part Two, the tendency to view law as an adjunct of sovereignty and centralised coercion leads him to displace or expel law from any significant role in modern forms of domination.

Conclusion: the problem of modernity

All important intellectual work is an engagement with the present and all significant contributions exhibit a deep ambiguity about the present with which they are trying to come to grips. In modern writing all significant contributions are both repelled and attracted by modernity (Berman 1982). One way of coming to grips with Foucault is to view him as deeply concerned to understand the trajectories of late twentieth-century life. Yet on the surface his concerns appear to be characteristically 'historical'. Most of his major studies are preoccupied with transformations occurring towards the end of the eighteenth century and the early nineteenth century. In his last major project, on the history of sexuality, his immediate focus of attention is even earlier, on the sexual ethics of classical Greece and Rome. The further back his immediate object of inquiry is located the more fiercely are his concerns located in the present.

Perhaps the most significant reason for the wide-ranging interest and engagement with Foucault's work is not so much that there is agreement with either his methods or his conclusions, but rather that his writings capture a deep and pervasive disenchantment with the modern condition. Gone is the optimism generated in the eighteenth century with the advance of reason; he problematises the idea that things get better, humanity progresses, becomes more civilised, dispenses with myth, superstition and religion as knowledge spreads its revelatory light over more topics of human concern. Similarly the projects of material and scientific progress of the nineteenth century and the expansionary vistas of the early twentieth century came increasingly under scrutiny.

The decades approaching the end of the twentieth century are marked by an escalating sense of rupture, the sense of the end of an epoch. The two great competing systems spawned by industrial capitalism, socialism and liberalism seem exhausted, the confrontation between the politics of left and right has become increasingly sterile. Part of the importance of Foucault undoubtedly stems from the fact that he captures the doubts and uncertainties that are so widespread today. Sure, technical advances of enormous potential continue to be produced, but science and technology are, for very good reasons, now viewed as just as much the illness as the cure. Foucault is a good example of a late twentieth-century thinker forced to abandon the optimistic scenarios of the nineteenth and early twentieth centuries. In rejecting the now tarnished optimism with its confidence in the exponential growth of knowledge and material progress there is always the danger of renouncing any concern with truth and knowledge and lapsing into pessimistic fatalism. Foucault walked this wire; his writings hover between an enthusiasm for the possibility of new emancipatory projects and a fatalism which sees all new quests for knowledge as yielding new and ever more sophisticated mechanisms of domination. He also resonates with the current mood in that he is short on policy and prescription since many of the problems we confront appear to be intractable. He is deeply hostile to all myths and utopias whether old or new, left or right. It is thus not surprising that there has been controversy over whether he should be regarded as optimist or pessimist, conservative or anarchist. We decline to enter these controversies because they are more about the politics of the commentators than they are about Foucault's politics. In so far as it is relevant, his own self-conception aligns him with what used to be called 'progressive' causes, but he refuses to use such a term because the very idea of 'progress' is one of the myths that Foucault's work seeks to problematise.

We suggest that there is no embarrassment in holding that some of Foucault's own political stances, such as his naive 'abolitionist' views about criminal justice or his ill-advised enthusiasm for the regime of the mullahs in Iran, are frankly silly and barely worth debating. Certainly he holds on to the enthusiasms of Parisian radicalism of the spring and summer of 1968 longer than most. However, any serious assessment of Foucault depends not on the causes he espouses but on what those who read him can do with his enormously fertile leads and suggestions. It is to this task that we now turn our attention.

Part Two
Foucault on Law

2
Law and Modernity

Introduction

We start our examination of Foucault's treatment of law by reminding the reader that we make no claim to 'discover' a ready-made theory of law in Foucault's writing. Law is never one of his explicit objects of inquiry. Nevertheless he has a considerable amount to say about law. The question of law not only figures significantly, but persistently returns in his texts.

Before exploring Foucault's treatment of law one qualification is needed. For convenience we refer throughout to the terms 'law' and 'the law' in the singular. However, there is an important sense in which such usage is misleading. Law is not and never has been a unitary phenomenon, even though the assumption that it is, has played a central role in most legal discourses and theories of law. We adhere to a view that law is a complex of practices, discourses and institutions. Over this plurality of legal forms 'state law' persistently, but never with complete success, seeks to impose a unity. This approach can be identified by the label 'legal pluralism'. To speak of 'law' or 'the law', as we do, can be excused only because it is easier for authors and reader alike; we return to this issue in Part Three.

Our account of Foucault's treatment of law is organised as follows: the main question we address is, what is the connection, if any, between law and modernity? We start with the most prominent and persistent theme in Foucault's writing, which we call 'law versus discipline'. We then criticise and assess the group of ideas associated with this position. Then we turn our attention to some other themes that are present in his work but which are generally less developed and have received less attention than the law versus discipline thesis.

Law comes to the fore in a group of Foucault's major texts written around the late 1970s. Law forms a significant motif in two of his most important and best-known texts, *Discipline and Punish* (D&P 1977) and *The History of Sexuality* (HoS 1978). Law also figures significantly in a group of essays and interviews from the mid 1970s collected under the significant title *Power/Knowledge* (P/K 1980). In this collection the second of his 'Two Lectures' is centrally preoccupied with the distinction he wishes to sustain between 'law' and 'discipline'. This lecture is important also in offering perhaps the most self-reflective overview of his work during this period. He speaks about his 'guiding principle' and his 'methodological imperatives' (TL 1980: 94). Foucault, the otherwise sceptical critic of all metatheories, comes closer in this text than anywhere else to laying out an overarching conceptual framework and philosophy of history. Because this text deals extensively with law and Foucault's central methodological and theoretical concerns we feel justified in attaching considerable weight to it.

Power and law

There is a general and persistent theme that runs through Foucault's texts. He describes the general trajectory of his engagement with the link between law and power in the following terms:

> It was a matter not of studying the theory of penal law in itself, or the evolution of such and such penal institution, but of analyzing the formation of a certain 'punitive rationality' ... Instead of seeking the explanation in a general conception of the Law, or in the evolving modes of industrial production ... it seemed to me far wiser to look at the workings of Power. (in Rabinow 1984: 337–8)[1]

Time and again he links law with the negative conception of power from which he strives elsewhere to escape, a 'juridico-discursive' conception of power, by which phrase he identifies all forms of law which specify prohibitions, 'Thou shalt not ...'.

> We shall try to rid ourselves of a juridical and negative representation of power, and cease to conceive of it in terms of law, prohibition, liberty and sovereignty ... We must at the same time conceive of sex without the law, and power without the king. (HoS 1978: 90–1)
>
> In short, it is a question of orienting ourselves to a conception of power that *replaces the privilege of the law* with the viewpoint of the objective, the privilege of prohibition with the viewpoint of tactical efficacy, the privilege of sovereignty with the analysis of a multiple and mobile field of force relations, wherein far-reaching, but never completely stable, effects of domination are produced. The *strategical model* rather than the *model based on law*. (TL 1980: 102; emphasis added)[2]

Since these remarks go to the very core of his treatment of law it is as well to offer a commentary on them. Here he sets up a concern to challenge the conventional, and in particular the Marxist, conceptions of state and power. His concern is to displace the equation of power with repression exercised by some unitary agency (whether the state or a ruling class). He points the finger at conventional conceptions of law, typically defined, as rules commanding behaviour backed by the threat of coercive sanctions; such a view he treats as an exemplar of precisely that view of power and the state which he seeks not only to avoid, but to transcend. This view of law has indeed played a central part in the history of legal thought. The formula 'law = rules + sanctions' lies at the heart of the positivist tradition in jurisprudence. It is thus pertinent that Foucault chooses to direct his fire against this widely held and influential conception of law, but fails to engage with any more sophisticated conceptions which seek to explore the connection between legal regulation, legal rights and constitutionalism in the bourgeois democratic societies.

It should be noted that in the remarks quoted above Foucault is concerned with law only illustratively; this is one important consequence of the fact that law itself was not his immediate object of inquiry. As a result he is not concerned to advance a more adequate conception of law. What Foucault does focus his attention on is the elaboration of an alternative conception of power. As we saw in Part One, one easy way of signifying this change is the contention that power is not simply negative but is also positive. Care needs to be taken, remember, not to interpret this move on Foucault's part as saying that power is in some sense 'good' rather than 'bad'. This is far from his intention. Rather what he wants us to attend to is that power exists in a multiplicity of different forms most of which do not manifest themselves in coercion. The forms of power to which he directs our attention are the 'small' powers. He underlines this point by persistently focusing attention on the power dimension of knowledge, captured in the linking achieved in his phrase 'power/knowledge'. The importance of his rejection of law as 'rules backed by sanctions' will become clearer when we consider the contrast he sets up between law and discipline.

Law-as-sovereignty not only manifests the coercive power of the state epitomised in the bloody vengeance wreaked on the bodies of those who offend against the king, but law also involves a distinctive production of truth. Not only do the procedures of law (trial, cross-examination, etc.) provide authorised means by which the truth is discovered, but once enunciated law provides the guarantee of this truth. Against law's self-image Foucault insists:

> Law is neither the truth of power nor its alibi. It is an instrument of power which is at once complex and partial. The form of law with its effects of prohibition needs to be resituated among a number of other, non-juridical mechanisms. (P/K 1980: 141)

Not only does law exhibit its own 'will to truth', it declares the guilt of offenders, and the truth/validity of its own rules, but it exerts 'pressure', 'a power of constraint', on other discourses: 'it is as if even the word of the law could no longer be authorized, in our society, except by a discourse of truth' (OoD 1981: 55). Indeed in modernity law, along with science, provides the privileged source of truth: 'It's the characteristic of our Western societies that the language of power is law, not magic, religion, or anything else' (P/K 1980: 201).

The truth of law is inscribed in its ultimate capacity to impose violence. Relations of domination are imposed, Foucault suggests, in a passage with strong Nietzschean overtones, through its 'rituals, in meticulous procedures that impose rights and obligations':

> the law is a calculated and relentless pleasure, delight in the promised blood, which permits the perpetual instigation of new dominations and the staging of meticulously repeated scenes of violence. The desire for peace, the serenity of compromise, and the tacit acceptance of the law, far from representing a major moral conversion or a utilitarian calculation that gave rise to the law, are but its result and, in point of fact, its perversion ... Humanity does not gradually progress from combat to combat until it arrives at universal reciprocity, where the rule of law finally replaces warfare; humanity installs each of its violences in a system of rules and thus proceeds from domination to domination. (NGH 1984: 85)

The history of legal truth reveals the shifting dependence of legal thinking on other systems of knowledge. From the mid nineteenth century legal thought became suffused with elements drawn from the psychological sciences, most evident in the abiding concern to understand the 'dangerous individual'; these developments were born at the boundaries and interchanges between law, psychiatry, psychology and medicine.

> Only an act, defined by law ... can result in a sanction ... But by bringing to the fore not only the criminal as author of the act, but also the dangerous individual as potential source of acts, does one not give society rights over the individual based on what he is? (1988a: 150)

The full implications of this shift have been held back and impeded by the continuing concern with legal issues such as intention, motive, responsibility and the elaboration of defences. Foucault says, somewhat cryptically, that the delay in fully grasping the 'new' principle

> indicates a foreboding of the dreadful dangers inherent in authorizing the law to intervene against individuals because of what they are; a horrifying society could emerge from that. (1988a: 151)

Thus while Foucault is concerned to mark a distinction between law and power, nevertheless in his detailed analyses law is itself a dangerous and problematic manifestation of power.

Sovereignty and right

Foucault's application of the conventional imperative view of law as the commands of a political sovereign is made more concrete when he articulates a more specifically historical thesis. In his overview of the rise of modernity his starting point is the 'Classical Age' (stretching from the late sixteenth century through to the second half of the eighteenth). His purpose is to treat the formation of the great centralised nation states, the paradigms for which were Britain and France, as the precursors of the distinctively modern social formations. He draws attention to a significant historical reversal; in the classical age trials were secret while punishment was public, but in the modern age trials are public, but punishment is, if not secret, then withdrawn behind institutional walls. The main characteristic of the classical age he identifies as the monarchical state constructed around the integration of 'law' and 'sovereignty'.

> Law was not simply a weapon skillfully wielded by monarchs: it was the monarchic system's mode of manifestation and the form of its acceptability. In Western societies since the Middle Ages, the exercise of power has always been formulated in terms of law. (HoS 1978: 87)

> Thus we see inscribed in the institutions of absolute monarchy ... the great bourgeois, and soon republican, idea that virtue, too, is an affair of the state, that decrees can be published to make it flourish, that an authority can be established to make sure it is respected. (M&C 1965: 61)[3]

His most distinctive thesis is that it was not merely that law and sovereignty constituted this 'juridical monarchy', but that law and sovereignty have remained central to the self-understanding of the modern forms of state power despite the passing of monarchical rule, extinguished either by revolution or absorbed into constitutional monarchies. His point is that the persistent focus on sovereignty and centralised law obscures the most distinctive features of modernity, specifically, the key importance of disciplinary power. It is an important part of his project to insist that, while contained within the shell of the old monarchical forms, modernity radically differs from the classical age. Hence his call to 'cut off the head of the King'.

> At bottom, despite the differences in epochs and objectives, the representation of power has remained under the spell of monarchy. In political thought and analysis we *still have not cut off the head of the king*. Hence the importance that the theory of power gives to the problem of right and violence, law and illegality, freedom and will, and especially the state and sovereignty ... To conceive of power on the basis of these problems is to conceive it in terms of a historical form that is characteristic of our societies: *the juridical monarchy*. (HoS 1978: 88–9; emphasis added)

It should be noted that what Foucault does here is to set out a very distinctive way of thinking about the place of law in modernity. When he refers to the persistence in so much modern political and legal thought of the central figure of 'the juridical monarchy' he is not thinking literally of 'monarchs' but more broadly includes all forms of unitary constitutionalism whether couched in terms of 'the Crown', 'Parliamentary sovereignty', 'the President' or 'the Constitution'. This way of thinking about the modern forms of political power he views as obscuring our understanding of the core features of modernity.

Foucault's line of thought seeks to undercut the self-evident and taken-for-granted discourses of modern constitutional thought. It has far-reaching implications for the way in which he addresses the place of law in modernity. The core of his thesis presents state legal systems as irrevocably linked to notions of sovereignty and his broad conception of 'juridical monarchy'. Thought of in this way, he presents law as being essentially *premodern*. We argue that, while Foucault is correct in stressing the inadequacy of political thought couched in terms of sovereignty to understand the complexity and diversity of modern power, he is wrong mechanically to equate modern state law with these discursive forms.

This account of the classical era characterised by the three elements of law, monarchy and sovereignty is supplemented by a conception of 'right' conceived as providing the discursive cement of the premodern era.

> The essential role of the theory of right, from medieval times onwards, was to fix the legitimacy of power; that is the major problem around which the whole theory of right and sovereignty is organized ... My general project has been ... to show the extent to which, and the forms in which, right (not simply the laws but the whole complex of apparatuses, institutions and regulations respon-

> sible for their application) transmits and puts in motion relations that are not relations of sovereignty, but of domination. (TL 1980: 95–6)

This conception of 'right' is rooted in notions of a 'divine right of kings' and in an imperative conception of law; the king's right is viewed as a right to command. But in the passage quoted above Foucault effects an unexplored shift; a shift from 'right' to 'rights'. This has the effect of locating the modern discourses of 'rights' (whether of private rights or human rights) as synonymous with the imperative notion of 'right' of the juridical monarchy and to bind them close by attributing to rights a general function of legitimation. This slippage from 'right' to 'rights' has serious consequences; it leads Foucault, like many other recent radical thinkers, to disparage the transformatory capacity of rights within modern political systems. It also produces a very distorted account of modern disciplinary society because he takes no account of the struggles for civil and political rights which have traversed all the fields of Foucault's own studies, whether of prisons, mental institutions or hospitals. He makes an important point when he insists that rights and freedoms are practices and no constitution or bill of rights can strictly 'guarantee' them, but it does not follow that such provisions are mere rhetorical flourishes.

> Liberty is a *practice* ... The liberty of men is never assured by the institutions of law that are intended to guarantee them. This is why almost all of these laws and institutions are quite capable of being turned around. Not because they are ambiguous, but simply because 'liberty' is what must be exercised ... I think it can never be inherent in the structure of things to guarantee the exercise of freedom. The guarantee of freedom is freedom. (Rabinow 1984: 245)

The questions of sovereignty and power connect with his pervasive concern with the enigma of modern power. In the classical era power was transparent, epitomised by the command-power of the king, while in modern society power has become diffused and its location becomes almost mysterious. This shift is epitomised in the visibility of political power and the often veiled reality of economic power. The result is that the tracking of domination, its strategies, techniques and technologies has come to form a central concern of both scholarship and political practice.

> The problem for me is how to avoid this question, central to the theme of right, regarding sovereignty and the obedience of individual subjects in order that I may substitute the problem of domination and subjugation for that of sovereignty and obedience. (TL 1980: 96)

In brief, Foucault is concerned to distinguish between the general characteristics of power in absolutist monarchies and liberal states. Despite his concern to distance himself from Marxism it is noticeable that here he comes close to returning to Marxism's concern with the mechanisms of general forms of domination. This tendency manifests itself in his treatment of discipline as a defining characteristic of modernity.

Discipline and law

A distinctive feature of Foucault's treatment of modernity is the importance he attaches to the emergence of 'the disciplines' as the characteristic and pervasive forms of modern power. Discipline and law are presented as dual but opposing processes. In the first place 'the effective mechanisms of power function in opposition to the formal framework that it had acquired' (D&P 1977: 222). In other words, while such disciplinary institutions as the prison are located within a juridical framework, the way they work is through disciplines that operate outside or in parallel with this legal framework:

> the disciplines characterize, classify, specialize ... they effect a suspension of the law that is never total ... Regular and institutional as it may be, the discipline, in its mechanism, is a 'counter-law'. (D&P 1977: 223)

The disciplines are to be found 'on the underside of the law' (D&P 1977: 223). The result is that all the institutions of incarceration, prisons, asylums and, by extension, factories and schools, operate in such a way as to 'naturalise' the legal power to punish at the same time as they 'legalise' the technical power to discipline (D&P 1977: 303). This produces the 'great carceral continuum', which functions through an ongoing 'communication between the power of discipline and the power of law' (D&P 1977: 304).

The techniques of surveillance are the epitome of disciplinary power. The nineteenth-century asylums for the insane were built on the dual pillars of 'surveillance' and 'judgement'. The asylum was a 'juridical microcosm' (M&C 1965: 251, 265).

> The asylum as a juridical instance recognized no other. It judged immediately and without appeal. It possessed its own instruments of punishment, and used them as it saw fit. (M&C 1965: 266)

The great institutions of incarceration, the asylum and the prison, epitomise a mutual dependence of law and discipline. Increasingly they become disaggregated and the disciplines continue to multiply and to operate independently of law. From the non-voluntary subjection to the 'gaze' of authority in the prison cell or hospital ward, the general theme of surveillance has become one of widening ramification as both human and technological surveillance, whether through direct observation (the reports of teachers, social workers, etc.) or by means of indirect scrutiny (the cumulative records of tax collectors, credit card operators, etc.). Increasingly surveillance expands through technological applications, for example through computerised and video tracking and scrutiny which has supplemented and displaced the 'gaze' of authority (Dandeker 1991). In drawing attention to the expansion of surveillance, care should be taken not to fall into some version of 'conspiracy theory' which views such expansions of power as irreversible and unchallengeable. Foucault's thematic of power should not lead to making a 'fetishism of power' in which 'power' becomes the subject of history, seen as undergoing evolutionary natural selection to produce ever more efficient and ensnaring forms (O'Malley 1993).

While Foucault sometimes counterposes 'law' and 'discipline' in order to highlight the distinctiveness of the modern disciplines, he is best understood, as is clear in the passages quoted above, as drawing attention to the interaction and interdependence of disciplinary practices and their legal framework. While he does not develop this insight in detail it is one which is familiar from the sociology of organisations, epitomised in studies of the police, where the formal and informal modes of operation can be seen to operate side by side. In addition, Foucault advances the interesting idea that law comes to be 'colonised' by the new disciplines being invaded by practices of observation and training, as is readily evident in contemporary juvenile justice (D&P 1977: 170).

Foucault should be understood as advancing a broad historical thesis. In a key passage he presents the rise of disciplinary administration as the 'dark side' of law.

> Historically, the process by which the bourgeoisie became in the course of the eighteenth century the politically dominant class was masked by the establishment of an explicit, coded and formally egalitarian juridical framework, made possible by the organization of a parliamentary, representative regime. But the development and generalization of disciplinary mechanisms constituted the other, dark side of these processes. The general juridical form that guaranteed a system of rights that were egalitarian in principle was supported by these tiny, everyday, physical mechanisms, by all those systems of micro-power that are essentially non-egalitarian and asymmetrical that we call the disciplines ... The real, corporal disciplines

> constituted the foundation of the formal, juridical liberties. (D&P 1977: 222)

In other words the advent of representative democracy existed side by side with the rise of an expanding disciplinary continuum which produced strong though barely visible mechanisms of domination. The practical reality of such subordination is itself legitimised by the juridical framework within which the disciplines are constituted; law functions as the mask of real power.

Alongside this vision of law as masking domination and as a vehicle of legitimation there is a trace in his writing that offers a rather different and more interesting view of law as, in some important sense, constitutive of the new forms of modern power:

> after the eighteenth century, the religious framework of those rules [of sexual morality] disappears in part, and then between a medical or scientific approach and a juridical framework there was competition, with no resolution. (GoE 1984: 357)

Law in this guise expresses the paradox of modernity. Confronted by the rise of the new disciplines, that are themselves exterior to law, the response of law is to seek to control or 'recode them in the form of law' (HoS 1978: 109). Such processes have been aptly described as 'juridification', that is that non-legal forms of discipline acquire legalistic characteristics.[4] Institutional rules governing such matters as work discipline or 'codes of conduct' not only come to be couched in legal language but invoke legalistic procedures, such as the right of representation and rights of appeal. Thus with respect to work-discipline employees are no longer subject to the traditional prerogative of 'hire and fire'; today 'charges' have to be laid in writing, the accused is given the right of representation and of appeal. The significance of the advancing legalism that lies at the heart of juridification is that it is a small step before the juridified processes come under the jurisdiction of state-law institutions, or before parallel quasi-judicial institutions are invented.

Yet despite this suggested 'recoding' of the new disciplinary mechanisms of power, Foucault is reluctant to grant any significant effectivity to law. Time and again he retreats to formulations in which law is viewed as 'utterly incongruous with the new methods of power' that are 'not ensured by right but by technique, not by law but by normalization, not by punishment but by control' (HoS 1978: 89).

While there are points in his treatment that focus on the interdependence of law and discipline, Foucault's broad developmental thesis is one which refuses to accord any major role to legal regulation in creating the distinctive features of modernity. Thus there is a dual impulse that leads Foucault to marginalise the role played by law. The first, as we have seen, is his historical analysis of the central role played by law in constituting the pre-modern complex of monarchy-law-sovereignty. The second motive that leads him to displace law is methodological. It stems from the reversal he advocates for the study of power and involves a shift of emphasis from state-power to local or 'capillary-power'. He uses 'capillary' as a metaphor to illustrate the multitude of small intersecting mechanisms through which power passes in contrast to the heavy-hand of coercive power exemplified by the harsh punishments of the pre-modern era, such as capital punishment or transportation.

One further feature of Foucault's treatment of 'discipline' should be noted. In *Discipline and Punish* he focuses on 'the body' as the target of first punishment and then discipline, but the subsequent trajectory of his writing sees an expansion: first to focus on the 'soul' (the 'knowable man', the psyche, subjectivity, personality, etc.) as the target of disciplinary practices; and second, in a more radical shift, to focus on the 'government of the self'. In brief, there is a shift from discipline to self-discipline. This progression is accompanied by the rise of the psychological sciences and professions (Rose 1989). In this change of emphasis from external constraint to internal states of conscience there is a corresponding further disaggregation of law and discipline.

Norms and normalisation

The operation of the new disciplinary mechanisms of power is ensured not by law, but rather by normalisation. 'The discourse of discipline has nothing in common with that of law, rule, or sovereign will ... The code they come to define is not that of law but that of normalization' (TL 1980: 106). He suggests that ever since antiquity there has been a competition between 'bio-power' and juridical power, made manifest in the 'growing importance assumed by the action of the norm, at the expense of the juridical system of the law' (HoS 1978: 144).

A significant but elusive feature of his treatment of the rise of the disciplines is this emphasis he places on the role of 'norms'. Discipline, rather than being constituted by 'minor offences', is characteristically associated with 'norms', that is, with 'standards' that the subject of a discipline come to internalise or manifest in behaviour, for example standards of tidiness, punctuality, respectfulness, etc. These standards of proper conduct put into place a mode of regulation characterised by interventions designed to correct deviations and to secure compliance and conformity; normalisation is thus counterposed to his prohibition-punishment model of law. The 'norm' is also implicit in surveillance in that it provides the criteria that the gaze invokes (for example, deference is translated into a sequence of behavioural signs) and deviance involves the infraction of the norm. It is through the repetition of normative requirements that the 'normal' is constructed and thus discipline results in the securing of normalisation by embedding a pattern of norms disseminated throughout daily life and secured through surveillance.

For Foucault another important difference between the disciplines and law is marked by their characteristic forms of punishment. In the place of the corporeal punishments of the old criminal law, 'exercises' and the repetition of tasks characterise the disciplinary model of bio-power. In turn these new forms of sanctions exist side by side with a greatly extended range of 'rewards' (for example, ranks, grades, prizes, badges, privileges). Thus the contrast he draws is between 'standards', which lay down general targets or criteria of judgement (such as attitude, demeanour) and 'rules', the legal form which he continues to conceive of as negative and prescriptive. The new forms of power that he associates with the displacement of the juridical system of law by norms and normalisation involve

> a power whose tasks is to take charge of life needs continuous regulatory and corrective mechanisms ... Such a power has to qualify, measure, appraise, and hierarchise, rather than display itself in its murderous splendor. (HoS 1978: 144)

It is his conception of law as commands that leads Foucault not to attend to the normative dimension of legal rules. There is one sense, however, in which he does address law as normative, and that is as a system of universal norms. One dimension of the distinction between law and disciplines is that between universal norms and the particularity of the disciplines; while the juridical system defines juridical subjects according to universal norms, the disciplines characterise, classify, specialise. They suspend the law such that the discipline is a 'counter-law'. His point is not to explore the normativity of law but merely to contrast 'universal juridicism' with the specificity of disciplinary interventions (D&P 1977: 223).

As we have seen, much of Foucault's account invokes a general opposition between disciplinary and juridical power, but just occasionally he insists that this is not his intent. He denies that he counterposes a 'society of sovereignty' and a 'disciplinary society'. His positive thesis is that 'in reality we have a triangle; sovereignty-discipline-government, that has as its primary target the population and its essential mechanism apparatuses of security' (G 1979: 18–19). In another similar formulation Foucault insists that 'the powers of modern society are exercised through, on the basis of, and by virtue of, this very heterogeneity between a public right of sovereignty and a polymorphous disciplinary mechanism' (TL 1980: 106). But despite formulations of this type his predominant tendency is to counterpose the disciplines to law, that is he conceives the two as antithetical and 'absolutely incompatible' (TL 1980: 104). A survey of Foucault's varied formulations of the relationship between law and discipline reveals a preponderant tendency to oppose law and discipline.

What sense should be made of his determined separation of law and disciplines followed by partial retreats? In the first place, Foucault separates these concepts primarily to give emphasis to the multiple origin of the disciplines in dispersed social practices, in particular those emanating from the 'human sciences'. On the other hand, he recognises that law has not disappeared or withered away. Since law does not constitute his object of inquiry, he has no special interest in exploring one rather obvious, but nonetheless attractive, hypothesis, namely, that associated with the rise of the disciplines there has occurred a change in the form of law.

Foucault draws attention to one consequence of the extended reach of the modern disciplines as the expansion of the reach of judges and judging:

> the activity of judging has increased precisely to the extent that the normalising power has spread ... The judges of normality are present everywhere. We are in the society of the teacher-judge, the doctor-judge, the educator-judge, the 'social worker'-judge. (D&P 1977: 304)

The result is identified as a passage from an age of 'inquisitorial' justice to an age of 'examinatory justice' (D&P 1977: 305). He also gives an earlier example of the extended political role of doctors around the time of the French Revolution. In addition to being technicians of medicine, doctors came to play expanding judgemental roles, for example in the economic distribution of assistance, and a 'moral, quasi-judicial role in its attribution' as 'the guardian of public morals and public health alike' (BC 1973: 42). As normalisation extends both its targets and its instruments, so judging becomes transformed. Judges become the bearers of normalising power while at the same time the judicial role is taken on by these other agents who make up the complex of 'the judges of normality'.

Foucault's account of the difference between law and discipline is at its sharpest where he draws this contrast between universal law and 'counter-' or 'infra-law', involving an 'infra-' or 'micro-penality' that takes possession of an area left empty or never colonised by the law, providing regulation for diverse types of behaviour. These micro-penalties involve 'offences' such as lateness, untidiness, disobedience, insolence. His point is that these wrongs are, on the one hand, so trivial as to be beneath the attention of law but, on the other, are the very stuff and heart of the modern disciplines.

Law, government and governmentality

One of the most important stages in the development of Foucault's work is the shift from a focus on discipline to government. He describes the general development of his project in terms that are worth quoting at length.

> It was a matter not of studying the theory of penal law in itself, or the evolution of such and such penal institution, but of analyzing the formation of a certain 'punitive rationality' ... Instead of seeking the explanation in a general conception of the Law, or in the evolving modes of industrial production ... it seemed to me far wiser to look at the workings of Power ... I was concerned ... with the refinement, the elaboration and installation since the seventeenth century, of techniques of 'governing' individuals – that is, for 'guiding their conduct' – in domains as different as the school, the army, and the workshop. Accordingly, the analysis does not revolve around the general principle of Law or the myth of Power, but concerns itself with the complex and multiple practices of a 'governmentality' which presupposes, on the one hand, rational forms, technical procedures, instrumentations through which to operate and, on the other hand, strategic games which subject the power relations they are supposed to guarantee to instability and reversal. (in Rabinow 1984: 337–8)

Modernity for Foucault is marked by the emergence of 'government' and 'governmentality'. As we have seen he uses the term 'government' in a way that is very different from the conventional sense of state executives and legislatures. One important implication of Foucault's conception of government is that it is consistent with his downgrading of the importance of the state and with it legal regulation. We return below to this relative neglect of the state since it has important implications for his treatment of law.

The reason that Foucault's discussion of 'government' and 'police' is important for our present concern with his treatment of law is that time and again he stresses the essentially non-legal character of his expanded conception of government. He insists that 'government' is

> not a matter of imposing laws on men, but rather of disposing things, that is to say to employ tactics rather than laws, and if need be to use the laws themselves as tactics. (G 1979: 13)

And again:

> the instruments of government, instead of being laws, now come to be a range of multiform tactics. Within the perspective of government, *law is not what is important.* (G 1979: 13; emphasis added)[5]

The deployment of 'multiform tactics' is illustrated in the link that exists between 'government' and 'population' where a variety of experts (quantifying, calculating and codifying) scattered across a range of agencies generate social policies that operate both to constitute the 'social problems' at which governmental action is directed and actively to regulate, control and coordinate the targets thus created. Whether Foucault is correct in suggesting that in this context of modern government 'law is not important' we consider later.

Foucault abandons the historical distinction between the classical and modern period. In its place he adopts a set of historical stages that push the juridical state back into the feudal period, with the 'administrative state' grounded in 'regulation' emerging in the fifteenth century and the 'governmental state' in the seventeenth and eighteenth centuries (G 1979: 21). The governmental state is characterised by the importance of the themes he had previously announced, the central focus on the regulation of 'the population' (rather than territory) and the role of 'police'. He gives this stage a new designation when he speaks of society being controlled by apparatuses of 'security'. He tries to capture the new form of government by speaking of 'the governmentalisation of the state' by which he seeks to embrace the whole range of governmental activity, the 'multiform tactics'.

A significant implication of this treatment is that it amounts to a tacit renunciation of the view that absolutism marks a more or less sharp transition to the modern forms of power that he had previously encompassed within the 'disciplinary society'. Now there is no break or sharp transition, but rather an expansion in the range and scope of governmental institutions. But it is of the greatest significance that what gets missed in this protracted process, stretching from the eighteenth century to the present, is any attention to the democratisation of the representative institutions and, more generally, with forms of participation in governmental processes, whether it be the rise of political parties or participatory organisations such as trade unions. One further consequence is that there is no place in his treatment for the notion of citizenship and certainly nothing which corresponds to any idea of an expanded citizenship that moves from the formal civil rights of the eighteenth century to the securing of universal franchise[6] and the social citizenship, epitomised by the welfare state, by the mid twentieth century (Marshall 1963).

Foucault's focus is upon the emergence of a concern with 'security' within modern governmental rationality. The English word 'security' does not convey the sense of Foucault's discussion; the term 'welfare' is probably closer. It is embedded in the shift from a view of individuals as 'subjects' to one in which they are conceived as the bearers of 'interests', that is, they are economic subjects, subjects or, to be more precise, 'subjects of the state', considered only in so far as the state requires to regulate their conduct or to demand performance from them, military conscription and imposition of taxes being two central examples. The individual, when considered as a bearer of interests, requires the state to take cognisance of those interests, in their multiplicity and complexity. 'Security' functions not by negative prescription or refusal, but rather through the specification of a range of tolerable variation. Thus liberalism constructs a complex governance within which political, economic and juridical instances of subjectivity are dispersed.

The association of 'security' with 'liberty' marks not merely the rise to prominence of rights discourses, but involves the idea that the systematic realisation of political and juridical rights are essential conditions of 'good government' which is itself a precondition for the persistence, stability and prosperity of both economic and political government. Gordon succinctly captures this governmental role of rights: 'disrespect of liberty is not simply an illegitimate violation of rights, but an ignorance of how to govern' (Gordon 1991: 20). It needs to be stressed that Gordon goes significantly beyond Foucault in giving weight to the juridical forms of liberty. This concern with rights and the conditions of prosperity can be seen as reaching a high point in Keynesian economic strategy in which the attempt to master cyclical economic crises and to secure the emblematic goal of full employment are conceived as preconditions of both economic prosperity and political stability. The 'security-liberty' characterisation of liberalism poses the question: *what part does law play in modern governmental rationality?*

Recall that Foucault asserts 'law is not important' (G 1979: 13). His account of the place of law is, in fact, more developed. The transition to modern governmental rationality involves a distinct and significant shift from some of his earlier positions associated with his expulsion of law discussed above. As has been demonstrated, in his major texts produced between 1975 and 1977, *Discipline and Punish, History of Sexuality* and *Power/Knowledge,* he equates law with sovereignty and the juridical monarchy; he is at pains to stress not only the dispersion but the privatisation of disciplinary power.

However there are in this group of texts hints of a different conceptualisation, one we can locate as the retreat from a transition from 'law to disciplines' to a new focus on 'law and regulation'. One of his key formulations posits this historical shift from law to regulation. This transition occurs when he suggests that law does not simply 'fade into the background' (HoS 1978: 144). It is this insight that is developed in the 'late Foucault', the Collège de France lectures of 1978 and 1979,[7] the essays 'Governmentality' (G 1979) and 'Omnes et Singulatim' (O&S 1981), the interest in 'liberalism' and, more generally, in his concern with the 'government of the self'. The root of this change of approach is the basic but important point that 'society' is an entity that had, during the course of the eighteenth and nineteenth centuries, to be discovered, whereas the nation conceived as a 'territory' was something that could be acted upon. But 'society' is a 'complex and independent reality that has its own laws' and thus cannot simply be acted upon (*Foucault Live* 1989: 261). Society necessitates 'good government', getting it right, since undesired results and unintended consequences of any active intervention may actually make things worse. It is this caution about the desirability and even possibility of government that sparks his interest in liberalism. In this phase of his work the earlier expulsion of law from modernity is significantly modified. Now his conception of law focuses on the purposive rationality of the legislative output of representative legislatures. He emphasises the increasing particularism of regulatory instruments. The previous conception of law as a totalising and transcendent unity is superseded by the historically specific production of regulatory devices that mediate between state and civil society and between state and individual. Foucault never developed this line of thought, but its presence underpins our claim that the study of the part played by law in modern governance is consistent with and can draw stimulus from Foucault's work.

Foucault's expulsion of law

We have demonstrated that Foucault's project of redirecting the study of power and of exploring the part played by the disciplines in modern government has as one of its distinctive effects the displacement of law.

> the analysis ... should not concern itself with the regulated and legitimate forms of power in their central locations ... On the contrary, it should be concerned with power at its extremities ... with those points where it becomes capillary ... one should try to locate power at the extreme points of its exercise, *where it is always less legal in character*. (HoS 1978: 96–7; emphasis added)

From this injunction to seek out power in its less legal manifestations it follows that research on the nature of power should be directed not towards the juridical edifice of sovereignty and the state apparatuses, but towards domination and the operations of power in their dispersed and localised sites. Thus it is apparent that the most distinctive features of Foucault's account of the historical emergence of modernity led him to present a view which can be aptly summarised as the expulsion of law from modernity. This 'expulsion of law' is found in his metahistorical thesis that law constituted the primary form of power in the classical or pre-modern era and in his point that law lingers on in the doctrine of sovereignty which continues to play a significant ideological role in political discourse. In the real world of power, law has been supplanted by the disciplines and by government as the key embodiments of power in modern society.

His expulsion of law is explicit in one of his most distinctive formulations.

> We must eschew the model of Leviathan in the study of power. We must escape from the limited field of juridical sovereignty and state institutions, and instead base our analysis of power on the study of the techniques and tactics of domination. (TL 1980: 102)

This methodological move presents Foucault with a serious difficulty: how to secure a focus on localised power without at the same time ignoring the indisputable significance of state and other forms of centralised and institutionalised power? As we have seen in Part One this is one of the most serious weaknesses in Foucault's work. We demonstrate in Part Three that it need not be a fatal weakness and that a theory of governance can adequately take account both of the diffusion of micro-powers and the aggregation of such powers at the level of the state and other institutional levels.

The destiny of law

As we have seen, Foucault's expulsion of law from modernity leaves law with, at best, an ambiguous role in his shifting conception of disciplinary and post-disciplinary society. He does, however, offer other suggestions as to the destiny of law: 'it is part of the destiny of the law to absorb little by little elements that are alien to it' (D&P 1977: 22). Foucault's account of the decline of law does not seem to involve the thesis that law will wither away. Rather his position can be characterised as allocating to law an increasingly subordinate or support role within contemporary disciplinary society.

> I do not mean to say that law fades into the background or that institutions of justice tend to disappear, but rather that the law operates more and more as a norm, and the judicial institution is increasingly incorporated into a continuum of apparatuses (medical, administrative, and so on) whose functions are for the most part regulatory. (HoS 1978: 144)

Here he suggests two distinct tendencies. The first tendency is a version of the widely held view that counterposes law and regulation, seeing the rise of administrative and technological regulation as signalling a decline or demise of law (Hayek 1982). What, for Foucault, distinguishes the new forms of power is that they function primarily by means of surveillance. For reasons that he does not explain he views law as not having the capacity 'for the codification of a continuous surveillance' (TL 1977: 104). This view does not strike us as persuasive since legal mechanisms are increasingly directed toward setting up the procedural mechanisms of a wide variety of systems of surveillance, for example by stipulating what records may be maintained by credit agencies, what must be disclosed to those subject to surveillance, and to whom such information may be disclosed. In short, modern law increasingly functions through such 'once removed' forms of regulation.

In general, Foucault's image of modern law is one of a mechanism that is ineffectual and generally epiphenomenal, confined mainly to providing legitimations for the disciplinary technologies and normalising practices established by other mechanisms (D&P 1977: 222). The kind of evidence on which he relies is illustrated by his contention that while legal principles have increasingly sought to restrict the incarceration of juveniles without explicit legal authorisation, yet incarceration continues to increase under the auspices of the new disciplines of medicine, social work and psychiatry. But the question that needs to be posed is: what does the evident and undisputable fact that law is relatively ineffective demonstrate? It is too weak merely to reply that law is ineffective. What we need to ask is: how is law implicated within different social relations? In particular we need to direct our attention to some of the persistent questions of classical jurisprudence about the capacity of legal control of diverse social practices, whether of the 'new' disciplines or the 'old' disciplines. The adequate pursuit of these questions requires the abandonment of jurisprudence's assumption of legal effectiveness, but it also requires the rejection of Foucault's presumption of legal weakness. Again we need to recognise that fruitful inquiry requires us to focus on the interaction of law and other disciplinary practices. Only then is it possible to distinguish between the different forms of legal effectiveness and to explore our intuitive judgement that there is considerable variation in the effectivity of law. This allows us to return to those classical but still important questions about the capacity of law to check non-legal forms of power.

One persistent feature of Foucault's reflections on the destiny of law is the contention that there is some fundamental incompatibility or tension between the legal form and the new forms of power.

> For this is the paradox of a society which, from the eighteenth century to the present, has created so many technologies of power that are foreign to the concept of law: it fears the effects and proliferations of those technologies and attempts to recode them in forms of law. (HoS 1978: 109)

> We have been engaged for centuries in a type of society in which the juridical is increasingly incapable of coding power, of serving as its system of representation. Our historical gradient carries us further and further away from a reign of law that had already begun to recede into the past at a time when the French Revolution and the accompanying age of constitutions and codes seemed to destine it for a future that was at hand. (HoS 1978: 89)

This emphatic position receives a characteristically Foucaultian treatment. Over time he offers a series of alternative formulations without offering a more elaborated treatment. For example, he describes the fate of law in the following terms: 'the procedures of normalisation come to be ever more constantly engaged in the colonisation of those of law' (TL 1980: 107). His terminology changes, but formulations such as 'absorbing' (D&P 1977: 22), 'recoding' (HoS 1978: 109), 'incorporation' (HoS 1978: 144), 'colonisation' (TL 1980: 107), and 'reorganisation of right' (TL 1980: 107) are each attempts, none more satisfactory than the others, to specify the character of the engagement between law and the new disciplines.

Modern power characteristically employs the discourses of law, but Foucault is concerned to suggest that this is a surface phenomenon, even an ideological phenomenon, although as we have seen he avoids this term. His point, we suggest, is to insist on some basic incompatibility between the form of law and the new disciplines. This incompatibility arises only from his own insistence on the unbreakable link between law and sovereignty and the command model of law that it generates. The connection between law and the new disciplines is much less troubling and involves no necessary tension or contradiction the moment we abandon his unitary model of monarchical law. A more adequate conception of law must start by conceding that like all other social relations law is subject to both change and variation in form. Once this simple, but fundamental, point is made it allows us to re-present the problem in a more manageable and fruitful way. The question now becomes: how is law articulated with the new disciplines? Or, more accurately, how are the shifting forms of law articulated with the new disciplines? It is this reformulation of Foucault's problematic that makes possible the retrieval of law that we pursue in Part Three.

3
Critique of Foucault's Expulsion of Law

Introduction

The primary theme that emerges from Foucault's treatment of the origins of the modern state and disciplinary society is one which casts law in the role of a pre-modern harbinger of absolutism. This line of thought can only impede inquiry into the part played by law in the governance of modern social relations. It is necessary to set out the deficiencies of this position in order to clear the space to embark on just such an investigation. Foucault's tendency to marginalise law contrasts sharply with the major drift of twentieth-century thought that has invested law with an increasingly central role in modern society. Such views cover a wide spectrum of recent social and legal theory; a thesis that posits the increasing centrality of law can be attributed to such diverse figures as Weber (1954), Poulantzas (1978), Dworkin (1986) and Luhmann (1985), to name only a few. The diversity of the accounts that attribute increasing significance to law reveal the radical revisionism that typifies Foucault's work.

Foucault's imperative conception of law

It is in this light that we now criticise Foucault's general equation of law with pre-modern forms of power. His conception of law as the commands of a sovereign backed by sanctions imposed on the bodies of the transgressors corresponds to a somewhat simplistic, if albeit common, view of law which equates 'law' with the punitive forms of criminal law. He captures this imagery in his now famous description in the opening pages of *Discipline and Punish* of the fate of the regicide Damiens whose body is torn apart in slow and painful stages. Crimes were literally conceived as offences against the sovereign. The problem is that Foucault implies from the spectacular and atypical offence of regicide that this is all there was to law in the classical era, the period of the formation of the European nation-states. It is significant that Foucault entirely fails to consider the much more fundamental process involving the elaboration and systematisation of the 'everyday' property offences which took place during this period. This neglect of the core economic offences parallels his more general lack of attention to economic relations. Similar charges of omission and oversimplification can be made against his suggestion of a leap from torture to the rise of incarceration; Foucault neglects the long process of elaboration of the criminal trial and procedure (Langbein 1977; Minson 1985: 83–90).

The equation of law with commands tends to reinforce the commonplace reduction of law to criminal law. While criminal law lends itself to being viewed as orders backed by threats this has only ever been one of the faces of law. In partial defence of Foucault it should be noted that the dominant tradition of legal theory in the English-speaking tradition focused on exactly the same characterisation of law. John Austin, whose influence lasted into the mid twentieth century, defined law in just this way, as commands of the sovereign backed by the threat of sanctions (Austin 1955). The other faces of law which, in so far as one can safely quantify law, make up its great bulk of provisions concern the detail of economic and kinship relations and the distribution of social authority. Foucault's conception of law entirely ignores, eliminates, suppresses all of this as well as the great mass of criminal law and what has so misleadingly come to be called 'private law'.

This is exactly the point that H.L.A. Hart so famously made against John Austin in *The Concept of Law* (Hart 1961); an imperative conception of law simply omits too much. In addition it also imports a dangerously oversimplified history of law, one which views the development of law as exclusively the consequence of the centralisation of power through the monopolisation of the means of organised violence in the institution of the monarchical state. Again this view corresponds to a certain commonsense view of legal history. It is a view which is not entirely false; there is an important link between state formation and the expansion of state-law, but again it tells only a part of the story. The simple conclusion is that working, as he does, with such an inadequate conception of law it is predictable that many of his pronouncements about law contribute little to an investigation into the part played by law in the history of the changing forms of power and government.

Foucault's derivation of law from monarchical power eliminates a more adequate history of law as emanating from dispersed sites of royal power, popular self-regulation, customary rights, competing specialised jurisdictions (ecclesiastical, guild, commercial, etc.), local and regional autonomies, and other forms of law. It was within this reality that royal power, reaching its zenith in the absolutist state, fought a never ending and never entirely successful battle to subdue and unify. The equation of law with negative proscription involves the acceptance of an ideological conception that came to form the conventional view of the monolithic unity of state and law; a view that, paradoxically, Foucault is prepared to accept in launching his own critique of the presumption of a monolithic state power.

Sovereignty and rights in monarchical and liberal states

Foucault relies heavily on a rather primitive equation of sovereignty and absolutism. His conception is constructed from a literal connection between political sovereignty and juridical sovereignty; the medieval king was the sovereign.[1] As Habermas comments, Foucault leaves the 'ungrounded impression that the bourgeois constitutional state is a dysfunctional relic from the period of absolutism' (Habermas 1987a: 290). However, Foucault makes an important point when he draws attention to the persistence of the discourses of sovereignty which are captured in his stylish aphorism about our failure to cut off the king's head quoted above (HoS 1978: 88–9). The key issue is whether or not the modern discourses of sovereignty are so heavily imprinted with the legacy of absolutism that modern forms of sovereignty are unable to overcome these origins. More concretely the issue is whether the modern discourses of popular or democratic sovereignty can be anything other than a barely disguised form of absolutism. It is worth recalling Foucault's formulation of the issue: the constitutionalism of the modern period involves 'the establishment of an explicit, coded and formally egalitarian juridical framework, made possible by the organisation of a parliamentary, representative regime' (D&P 1977: 222). This juridical form involved a system of rights that are, in principle, democratic. However, and here is the key point, this constitutional edifice only serves to mask the

> tiny, everyday, physical mechanisms, by all those systems of micropower that are essentially non-egalitarian and asymmetrical that we call the disciplines ... The real, corporal disciplines constituted the foundation of the formal, juridical liberties. (D&P 1977: 222)

There are two aspects to Foucault's critique of constitutionalism. The first is that sovereignty remains a centralised power to command that is more or less impervious to the democratic discourses within which it is located. For example, the more or less rapid succession of the French Revolution by the Napoleonic monarchies or the more recent tendency of the major western democracies towards presidential rule with seriously weakened representative assemblies or parliaments provide the kind of evidence that might support this thesis of the persistence of classical sovereignty. The second strand of his critique treats constitutionalism as a largely ideological device; it purports to describe the location of power and control, while in fact the distinctively modern forms of domination are actually constructed on the basis of the less visible but pervasive disciplines.

We are confronted with one of the most fundamental questions of our epoch. Put in its simplest form, it is true that the modern constitutional democracies have rarely provided government for the people, and have never been government by the people. On the other hand the democratic form is more than a mask, it provides very real and tangible restraints on the exercise of power; but there can be no doubt that representative democracy – even at its worst and most corrupt – is a marked improvement over the absolute monarchies of the eighteenth century. Modern democracy and constitutionalism has to be approached as a dilemma, that is, with a genuine doubt about its achievements and its potential for realising participatory democracy. The deficiency in Foucault is not that he problematises constitutional democracy, but that his answer comes down so unambiguously on one side of the dilemma. The side of the paradox which Foucault omits is the extent to which the new forms of disciplinary power have already or can potentially become subject to processes of legal rights and legal regulation.

Foucault's linkage of law and sovereignty as a pre-modern form appears counter-intuitive since, on the face of it, the most significant forms of post-absolutist government seem to place heavy reliance on such features as 'the rule of law', 'the separation of powers' and other distinctively legal or constitutional features. This has led some commentators to offer an interpretation of Foucault which seeks to acquit him of any general equation of law and absolutism. Although the detail of their argument differs, François Ewald (1990) and Jerry Palmer and Frank Pearce (1983) draw a distinction between 'law' and 'juridical' such that discipline is not counterposed to 'law', but to the 'juridical' or to the neologism which Foucault introduces, the 'juridico-discursive'. The point that Foucault makes is that power acts by 'laying down the law', that is, through acts of discourse that 'speak' the rule (hence the 'discursive' dimension of his concept 'juridico-discursive') (HoS 1978: 83). Thus for Ewald and Palmer/Pearce, it is this discursive form that links law with sovereignty. We do not find this argument persuasive because, both ancient and modern, law has exhibited these discursive characteristics, for example, by the use of abstract 'legal subjects'. While the form of legal discourse is of undoubted significance, these features do not serve any useful role in distinguishing traditional and modern forms of law. Our suggestion is not only simpler, but more direct and to the point; it is to recognise that the strict association which Foucault makes between sovereignty and law is at best unhelpful and at worst simply perverse in denying the self-evident truth of the intimate connection between modern forms of power and legal mechanisms.

The same objection needs to be made with respect to the way in which Foucault ties legal rights to sovereignty. The serious flaw that follows directly from his conflation of monarchical 'right' and legal 'rights' is that he treats modern discourses and practices of rights as if they were nothing more than the repetition of the 'old' discourses of 'right'. Thus, while he refuses the concept ideology, it is precisely through this 'absent' category that he treats the modern liberal preoccupation with rights as ideological mystification or false consciousness. However, we cannot simply reverse Foucault and assume that all the modern discourses of rights provide a satisfactory description of actual restraints on power. We suggest that legal rights always exist in tension with the exercise of governmental power. Nothing guarantees that rights operate to constrain the exercise of power. It is merely that they provide a significant form of and forum for public argument. Litigation in courts of law that invokes rights to restrain governmental power is but the most obvious example. The field of the politics of rights also plays itself out in the wider field of public political controversies which occur in public arenas such as parliaments and the media. For example, Habermas provides a very important way of addressing such issues when he suggests that the existence of welfare rights exhibits an important paradox; on the one hand they hold out the legal means for securing the interests of welfare claimants while they can operate in such a way as to render claimants as powerless recipients of a paralysing paternalism (Habermas 1987b: 290–1).

It is not necessary to hold illusions that rights always secure the promises they hold out; that, for example, pay equity legislation guarantees equal pay for women. But what we might call 'realism about rights' or a healthy scepticism should not give rise to a denial of any potential political value to tactics that seek to invoke rights against the incursions of disciplinary power and to advance or expand new rights. We think that Foucault is wrong in suggesting that the politics of rights leads only into a 'blind alley':

> it is not through recourse to sovereignty against discipline that the effects of disciplinary power can be limited, because sovereignty and disciplinary mechanisms are two absolutely integral constituents of the general mechanism of power in our society. (TL 1980: 108)

The odd feature about this formulation is that while elsewhere he distinguishes sovereignty and discipline as historically distinct modes of power, he now asserts that they are 'integral constituents'. The political implications of this position bear the imprint of an earlier phase during which, after 1968, he moved close to the Maoist student radicals. By 1972 he distinguishes himself from their political demands. It is, however, significant that he does so by adopting a position that is, if anything, to the 'left' of Maoism.[2] At the height of the post-1968 passions the Maoists called for the organisation of popular or people's courts to try the crimes of the powerful. Foucault disagrees with this demand, but he does so on the essentially anarchist grounds that any reliance on the form of the court must ensnare popular justice with features characteristic of bourgeois state institutions (P/K 1980: 1–36), in particular institutionalisation.

At the very end of the 'Two Lectures' he advances an alternative to reliance on sovereignty and the discourse of rights, namely the development of a 'non-disciplinary form of power':

> it is not towards the ancient right of sovereignty that one should turn, but towards the possibility of a *new form of right*, one which must indeed be anti-disciplinarian, but at the same time liberated from the principle of sovereignty. (TL 1980: 108; emphasis added)

Unfortunately, and possibly significantly, he says nothing about what this 'new form of right' might be. Nor does he ever return to explore this idea (Miller 1993: 292–3).

It may be too harsh to say that he never returns to this topic. Late in his work he did return to grapple with issues that touch on this range of issues, when he became interested in the distinctive characteristics of liberal government (Gordon 1991: 14–41). Foucault's remarks on what Colin Gordon helpfully calls 'real liberalism' are difficult to access because they are scattered and, in the main, drawn from lectures of which there is yet to be authorised publication. In his scattered comments Foucault is concerned to flesh out his earlier thoughts about the emergence of an expanded conception of police and its association with projects of 'population' concerning the numbers, health and prosperity of people who, in the early modern towns, were beginning to secure some of the attributes of citizenship. This sense of liberal government is a useful corrective to the equation of liberalism with state-abstentionism and *laissez-faire* economics. It does not, however, go very far towards identifying the characteristics of modern government or the role of law within it. At best we can view Foucault as looking in this direction of inquiry, despite the fact that much of his earlier reflections on law and government had blocked the possibility of such inquiry.

Beyond the disciplinary society

There is a deep ambiguity, maybe even a contradiction, between Foucault's stress on the productivity of power and his bleak imagery of an oppressive 'disciplinary society' in which the individual 'finds himself caught in a punishable, punishing universality' (D&P 1977: 178). This ambiguity is echoed in his stress on the negative, prohibitive visage of law and the negative productivity of disciplinary power. At the heart of Foucault's expulsion of law lies his concern, manifest in all his interventions about power, to identify the emergence of distinctively new forms of power that characterise modernity. His account of the transition to modernity posits a series of displacements, starting in the seventeenth century. He assumes that centralised monarchical power was consolidated during the course of the eighteenth century such that by the second half of the nineteenth century we find in place a distinctive new form of power, disciplinary power. In contrast to the unitary form of state power, this disciplinary power is constituted through the play and interplay of a plurality of disciplines.

In charting the rise of disciplinary society Foucault employs a tactic that is characteristic of much of his work. In order to clear the space for a new thesis he displaces or contradicts the existing common sense; a strategy epitomised in his rejection of the 'repressive hypothesis' in the history of sexuality (HoS 1978: 45ff). It is in this context that we should understand the way in which he counterposes disciplinary power to juridical power. Foucault describes a new type of power, one which cannot be encompassed within the discourses of sovereignty. Disciplinary power lies outside sovereignty and thus does not depend on the centralised power of the state. It is in this sense that he describes the disciplines as being a 'counter-law' which operates 'on the underside of law' (D&P 1977: 223).

It is from a concern to highlight the distinctiveness and novelty of the disciplines that Foucault is led to oppose them to law. But it needs to be firmly insisted that, contrary to Foucault, disciplinary power is not opposed to law, but rather that law has been a primary agent of the advance of new modalities of power, law constitutes distinctive features of their mode of operation. This is not the place to attempt a history of the link between law and the disciplines, but some features can be indicated. To take the important example of labour discipline, in the eighteenth and nineteenth centuries there was a complex interaction between the use of criminal law in the form of vagrancy laws and anti-trade union laws which coexisted with the abstentionist legal endorsement of the patriarchal powers of 'the masters'. During the nineteenth century the disciplinary powers of employers were reinforced through the imposition of restraints on some of the cruder forms of labour discipline, such as the truck system. The twentieth century has seen the growth of juridification through the proceduralisation of work discipline that operates alongside the Taylorist economic devices, such as the piece-work system, regulating the intensity and quality of work-activity. At every stage in the complex history of the regulation of labour, legal mechanisms and devices have operated alongside the disciplinary 'counter-law' or 'underside of law' that Foucault highlights (D&P 1977: 223).

Now that we have articulated a general criticism of Foucault's treatment of law and the opposition between law and discipline we can move on to consider some other important features of his treatment of law. While Foucault never retracts his opposition of discipline and law, he develops a number of strands which go some way toward the presentation of a more sophisticated view of the relationship between law and modernity. To these strands we now turn our attention.

There is no doubt that he makes an important point in proposing that regulation is not and never has been synonymous with or bounded by law. Yet he misses the more important point that state law is always involved with, if not preoccupied with, the task of either exercising control over or exempting from control the different forms of disciplinary regulation. For example, while historically law ceded family discipline to the patriarchal father, this site has more recently become a major field of regulatory contestation. A more adequate account starts from the idea that the whole field of social regulation involves an ongoing process of expansion and contraction of the sites of regulation and the advance or withdrawal of the different regulatory techniques. A more persuasive response to Foucault's account of law in late modernity suggests that the trajectory of law is far more complex than he is prepared to admit. A more adequate account needs to stress a persistent increase in the range, scope and detail of legal intervention that produces a general movement towards an expanding legalisation and juridification of social life. It is within this framework that the issue of the role of law, of its advances and retreats and the changes in its active forms, can be more rigorously posed than is allowed by Foucault's counterposing of law and discipline.

One fruitful dimension is Foucault's undeveloped suggestion that law functions increasingly as a 'norm'. François Ewald, as we have seen, provides an examination and extension of Foucault's thesis about this 'normativisation' of law (Ewald 1990: 138–61). The norm is to be distinguished from the rule; norms identify general standards, not in the sense of 'principles' or meta-rules, as used by Dworkin, but rather as a set of standards; perhaps one way of grasping this idea is to extend Foucault's own notion of a discursive formation, to say that what Ewald suggests is a 'normative formation'. Ewald illustrates this idea by exploring the popular contemporary theme that late modernity is characterised by the rise of the risk or insurance principle (Ewald 1991; O'Malley 1991, 1992; Simon 1987). Law enforcement authorities are no longer able to provide protection against, for example, petty housebreaking and vehicle theft, with the result that we have witnessed the phenomenon of insurance companies insisting on the installation of alarm systems in private homes as a condition of issuing or renewing insurance policies. The implication we are forced to accept is that housebreakers do not get apprehended, and that we must rely on monetary compensation via insurance claims. This process is, we suggest, not to be understood as an advancing normativisation of law, but rather as the emergence of a shifting 'limit of law' that involves a significant expansion of contractualism. The general evidence that Ewald advances is consistent with the element in Foucault that seems to suggest a 'displacement' of law. But Ewald's account fails to add either substance to or confirmation of a thesis about the advance of the 'norm'. Ewald may well succeed in capturing Foucault's intention, but if he does then he reveals the deficiency in the normativisation thesis. The contention that we are witnessing a pervasive enlargement of the insurance principle points not to a change in either the form or content of law but rather to a development that far from displacing law, serves both to transform and to expand its reach. Ewald provides an important instance of the contention that legal regulation cannot and should not be examined in isolation but through its complex interconnection with other regulatory techniques.

Legal regulation becomes more and more deeply involved as one component of the detailed governance of many forms of social relations and institutions. It acquires an increasingly particularistic character laying down detailed rules and procedures for a host of specialised areas of activity, for example in detailed provisions concerning welfare entitlements, construction standards, product safety, credit transactions, and so on. It should be noted that this expansion of law is significantly associated with trends toward greater 'proceduralisation' which, rather than setting positive rules to control activities, lays down procedures for how decisions are to be taken, for example specifying what interests are to be consulted. Habermas has recently gone so far as to suggest that proceduralisation is the most significant potential of law in contemporary conditions (Habermas 1993). Alongside these developments is the complex phenomenon of the advancing constitutionalisation of citizenship expressed in many and varied extensions of both the forms and types of entrenched rights that go far beyond the classical political and property rights of the constitutionalism of the eighteenth and nineteenth centuries (Marshall 1963; Somers 1993). Once we abandon Foucault's narrow conception of law as a system of commands or prohibitions we are able to discover not a separation or distance between discipline, normalisation and law, but rather an interplay or even interpenetration of law, normalisation and discipline. It then becomes possible to insert some substance into his undeveloped suggestion about 'the triangle' of sovereignty-discipline-government (G 1979: 18–19).

For Foucault and Ewald the articulation of a norm is the by-product of the routinisation of the normalisation process. The norm itself has no history. It is perhaps for this reason that Foucault ignores the normative content of legal rules; he remains content to treat them as negative prohibitions. Given the complex trajectory of normative conceptions of 'responsibility', 'guilt', 'fault', and the whole battery of legal concepts and their close association with criminal procedure and modes of proof, it seems almost perverse that Foucault should ignore the normalisation of law that occurs in what is probably close historical association with the rise of disciplinary normalisation. Foucault simply takes no account of the internal and substantive aspects of the development of legal thought (Habermas 1987a: 289). Perhaps this line of criticism should be made more severe by noting how strange it is that Foucault's perceptive analysis of the specialised professional discourses is never brought to bear on legal discourses and their surrounding procedures and practices.

Foucault's thesis concerning the rise of normalisation as a new form of power should not be allowed to obscure the long history of normalisation that predates the rise of the modern state. In pre-modern society wide arenas of conduct were subject to strict surveillance, discipline and legal control through the regulation of consumption and the ordering of appearance; these regulatory mechanisms were subsequently abandoned or defeated, but new sites of normalisation became the subject of contestation (Hunt 1994). These interventions attest to the pervasive role of normalisation and discipline in pre-modern society and an awareness of these interventions emerge in Foucault's late reflections on liberal government (G 1979; O&S 1981). Normalisation, discipline and regulation are not new and, of course, Foucault knows this perfectly well, but in the search for the distinctive novelty of the modern he tends to obscure our vision of the regulated pasts.

The really interesting questions that need to be posed concern the shifts and transformations marking the changes that characterise the genuinely novel mechanisms Foucault has done so much to chart. These issues provide fertile avenues for research because they involve complex articulations of self-control, confessional techniques, legal regulation and disciplinary processes. As well, significantly, these shifts involve different modalities of the participation of law in these processes. Foucault's overemphasis on the novelty of disciplinary and normalising power also creates the risk of undermining the significance of his own account of modernity. While he frequently returns to suggestive ideas about modernity understood as manifesting the reconstitution of elements already in place and developing interstitially in the old order, his tendency to succumb to the 'big sweep', epitomised by his counterposing of law-sovereignty to discipline-normalisation, does much to undermine this potential.

Perhaps the central deficiency that permeates Foucault's treatment of discipline is the lack of any explanatory mechanism whereby the dispersed and plural 'disciplines' are aggregated into his pessimistic and negative utopia of the Gulag which is 'disciplinary society'. To identify such a strategic unification suggests the existence of some far-sighted and malevolent 'ruling class', but this Foucault explicitly rejects as one of the major grounds for his refusal of Marxism. Yet without some mechanisms of unification or aggregation the dramatic imagery of the disciplinary society is undermined. The resort to this tactic on Foucault's part is a manifestation of the problems that are inherent in his general thesis of 'strategy without strategists'. It is only by invoking a globalised conception, the disciplinary society, that he is able to sustain the idea of the displacement of the law-sovereignty complex. But this globalisation comes into conflict with his pervasive and forceful insistence on the dispersal of the sites of power and the plurality of the disciplines. Do the modern disciplines and normalising practices have any unity? More importantly, if so, where does this unity come from? How are the mechanisms of normalisation orchestrated?

Foucault attempts to resolve this difficulty by an inspired but ultimately unsatisfactory move. He insinuates unification by positing a unifying mechanism whose manifestation takes two well-known forms – the 'gaze' and the Panopticon. Both of these devices are significant in that they imply a process of unification or centralisation of diverse practices without the need to posit a unifying agent. In *Discipline and Punish* the metaphor of the Panopticon plays this role of imposing unity in that he treats the Panopticon as the essence of modern disciplinary power. This claim lays itself open to two objections. The first is the sociological challenge that whatever significance we might be prepared to accord to the Panopticon (and in passing it may be suggested that he greatly overemphasises the historical significance of what was a largely unrealised project),[3] there is no evidence to suggest that the other forms of normalising discipline were motivated by the same global aspirations. It is similarly apparent that there is no unitary 'gaze'; for example, psychiatrists, social workers and prison guards all deploy distinct and fragmented gazes.

A second set of difficulties surrounds Foucault's persistent, but elusive, identification of discipline with surveillance and 'normalisation'. He paints a set of vignettes of the 'discovery' of new sites of disciplinary intervention. For example, he contends that a moral panic surrounded the 'discovery' of juvenile masturbation that brought into play a heightened level of surveillance over the bodies of children and an objectification of sexuality through moralising medical discourses. His texts are full of the 'discovery' of the multiplication of disciplines that have their beginnings in 'little places' and that extend their range of operation. Much of the reason for Foucault's very considerable influence is that – in the most general and untheorised sense – the picture he is taken to have painted is of ever-extending and ever more intrusive mechanisms of power that insert themselves into every nook and cranny of social and personal life. There is no doubt that this strikes a chord of intuitive recognition and resentment in the self-consciousness of the late twentieth century. Foucault's sketch of an expanding disciplinarity shares much with the currently popular theme of 'juridification' that gives a label to the process through which there has occurred a steady advance of legal intervention into ever more spheres of social life.[4]

Not only does this picture fuel the anti-authoritarianism of the left, the innumerable projects of escape from modernity, but also the projects of the escape from the constraints of over-regulation on the part of neo-liberal conservatives. Now, of course, this storyline is not new, but Foucault's version of the undramatic but cumulative impact of the disciplines connects with contemporary sensibilities. However much we may empathise with this general orientation, it should not blind us to the fact that Foucault is far from convincing in establishing the massification of the dispersed disciplines into 'the disciplinary society'.

Conclusion: the dilemma of freedom

Much of the important period when Foucault's work was preoccupied with the question of power is marked by a rather stark segregation between the pre-modern and the modern; it is this more than anything else that determined his treatment of law. His final texts, those that display his uncompleted history of sexuality, mark a shift to a concern with self-government, the 'techniques of the self', that go a long way to overcome his earlier schematic separation between the pre-modern and the modern (HoS 1985a, b). He formulates this issue in terms of people's engagment in processes through which they constitute their identity via what he calls 'ethical techniques of the self' that have developed from antiquity down to the present. If we are to take full account of the shift of attention that Foucault proposes towards the micro-physics of power one of its most important implications is that it disallows any assumption that these dispersed powers form any kind of unity.

On the contrary the serious pursuit of Foucault's concerns should lead us to pose a different set of questions. How should we approach the study of the forms of articulation of the dispersed disciplines? How are they combined, how are their competitions resolved? It is towards inquiries of this type that the genealogical method is directed. At first sight such a stance seems to suggest a move away from law and other institutionalised mechanisms. But the shift to the focus on government of the self and to ethics does not announce a retreat to the private realm or a shift from the social to psychological inquiries. In an important sense it is a return to what, throughout the modern period, has been the core question of law and legal discourses, namely, the link between government and freedom. Down this route lies a fruitful line of inquiry that focuses on the role of both law and state, unnecessarily marginalised by Foucault, with respect to the coordination and condensation of the forms of power. The state apparatuses and state law are continuously driven to pursue projects of the unification of power; the success of these projects is always partial, limited and incomplete; their failures serve as impulses to further projects of regulation. Law is an ever-present participant in this success and failure of governance. It appears both as the means of restraining and channelling the projects of governance, yet at the same time is one of the projects' means of existence.

This approach leads us to pose the question of law as always a dilemma, the dilemma of government and freedom. But dilemma does not imply opposition or mere dichotomy. Government is not opposed to freedom, just as freedom has never been the mere absence of government. Rather government and freedom pose a dilemma in that they presuppose the other while at the same time they threaten or challenge the other. One form in which the dilemmic character of law manifests itself is with respect to legal rights; rights cannot guarantee freedom, but freedom cannot be achieved without rights. This dilemma is never posed by Foucault, but it is one that is consistent with his concern, in the final stage of his writing, with the conditions for freedom. The paradox is that Foucault's general line of thought opens up these inquiries about law while much of his specific treatment of law seems to exclude them and deny their relevance.

Part Three
Deploying Foucault for a Sociology of Law as Governance

4
Governance and its Principles

Introduction

Our task in this chapter is straightforward: it is to offer a basic definition of governance and to elaborate its four principles. Our discussion covers the complexities of governance such that the reader is prepared for our new sociology of law as governance. We conclude the chapter with a brief discussion of one of the central complexities, the extremes of governance.

We do not spend much energy in our discussion relating our points directly to Foucault; we do so only occasionally. The Foucaultian influence on our treatment of governance should be clear from the preceding chapters. The Foucault who inspires this part of our book is the Foucault who is interested in government alongside power, the Foucault who uses the neologism 'governmentality' to capture the dramatic changes in techniques of government developed in the western world from the eighteenth century onwards. This may not be the most popular Foucault, but we take it to be the most rewarding Foucault for those, like ourselves, interested in new directions for the sociology of law.

We are inspired not just by Foucault's direct discussion of governmentality (G 1979), but also and more importantly by the work of others heavily influenced by Foucault's work on this notion which is contributing to a distinctive approach. The flavour of this approach is captured in the collection edited by Graham Burchell, Colin Gordon and Peter Miller, *The Foucault Effect: Studies in Governmentality* (1991). We make reference to several of the essays from this volume. We also direct the reader occasionally to the work of Nikolas Rose and Peter Miller (Rose and Miller 1992; Miller and Rose 1990).

We do not deal directly with the notion of governmentality in building our account of governance. Rather, we use governmentality as a resource for and background to our account. We offer a sketch of governmentality here to complement our earlier discussions such that we allow the reader some insight into the richness of the Foucaultian work in the area and a brief understanding of our resource pool. Our sketch is drawn from Foucault's seminal essay (G 1979) and from various other essays in *The Foucault Effect* (Burchell 1991).[1]

In simple terms, governmentality is the dramatic expansion in the scope of government, featuring an increase in the number and size of the governmental calculation mechanisms, which began about the middle of the eighteenth century and is still continuing. In this way, governmentality is about the growth of modern government and the growth of modern bureaucracies. As Gordon (1991) recognises, this is the moment where Foucault meets Weber.

This simple definition is useful up to a point but it does not capture enough of the subtlety of Foucault's concept. It does not, for example, allow us to follow closely Foucault's periodisation. While government and its mechanisms have indeed boomed from the eighteenth century onwards, this period is hardly unique in the history of widespread, sophisticated governmental techniques. Ancient Egypt, ancient Greece, ancient Rome and many examples from both the western and eastern worlds in the period from the fall of Rome to the middle of the eighteenth century all mark boom times for just such government; all these examples could be regarded as instances of governmentality were we to use only this simple definition.

To enhance this simple definition such that the nuances of Foucault's governmentality are more easily recognised, we suggest a series of interconnected definitions around the following themes: the emergence of the reason of state; the emergence of the problem of population; the birth of modern political economy; the move towards liberal securitisation; and the emergence of the human sciences as new mechanisms of calculation.

From the sixteenth century on, a variety of doctrines around *raison d'état* began to emerge; a series of doctrines which understood the operation of the state according to principles which were internal to the state itself and which had their own autonomy. Principles of government were no longer transcendent principles of an order of things guaranteed by God. Instead, the correct principles for the organisation of the state came to be seen as immanent; the strength, economic and military, of the state itself became the goal of, and justification for, state action. This reorganisation of knowledge had the consequence of creating a new set of problems for governments. Governments now had to decipher the mystery of the state and calculate the correct principles for its ordering.

One of the ways in which these new problems were addressed was in relation to population. For Foucault, the concept of population allowed the art of government to overcome the obstacle created by the emergence of the reason of state. Government came to be a means to an end in relation to population concerns: how to guarantee the health, wealth, happiness, longevity, and so on, of the population; a whole series of strategies, which elsewhere Foucault has termed bio-power or bio-politics, sought practical answers to such questions.

A new science arose which took this new entity as its object: population became the proper concern of political economy. By addressing population, governments were able to target each individual, as a part of the population; the family as the unit of analysis of the nation was now clearly insufficient. The regularities of the population, in terms of mortality rates, epidemics, and so forth, could not be understood as part of the economy of the family (the *oeconomy*) and the new political economy, which gradually replaced this *oeconomy*, recognised this. The family was still an important instrument of government, but it was now secondary to the master concept of population. The new political economy, which sought to promote the flow of government between individual, family and state used the concept of population as the primary means of recasting the art of government. This science dealt with the governmental imperatives which flow between state and individual, taking the family as an instrument in these strategies, rather than as a model for them.

The emergence of liberalism marked an important transition point for discourses around the art of government. We are especially interested in the emergence of the liberal idea of a society where the liberty of individuals was seen as being potentially guaranteed through security. The reorganisation of government within the constraints of security can be seen as an elaboration of the theme of the government of fortune, that is, the management of whatever life or fate delivers. A whole series of social technologies answered this need to govern liberty; these include the public welfare mechanisms of the late nineteenth and twentieth centuries and the modern welfare states (Burchell 1991; Defert 1991; Ewald 1991).

Finally, a particular series of formal human sciences provided points of articulation for governmentality: the rational economic man of economics, the rational autonomous subject of psychology, the autonomous social of sociology, all emerged at about the same time. In addition, straddling these sciences, the science of statistics expanded rapidly: a set of facts about the state was reformulated as a set of very specific understandings of population; a precise knowledge of birth, mortality, morbidity, longevity, health, illness, suicide, contributed to the possibility of installing a new governmental rationality.

This complex of definitions, we argue, allows a firmer grasp of the elusive theoretical instrument which is governmentality. Using this complex, Foucault's periodisation is more easily followed. On the one hand, Foucault points us towards an exponential growth in government, in the elaboration and extension of what it means to govern fortune. On the other hand, he points us towards a series of very specific historical techniques which are formative of our present.

As we suggested earlier, while our new sociology of law as governance is primarily Foucaultian in its inspiration and direction, we draw on other traditions in its formulation. All of these but one are indirect. The strongest indirect influences are Machiavellian political theory and Weberian sociology, though there are definitely traces of ethnomethodological sociology and Parsonian sociology to be found as well. The other tradition which directly influences our proposed new direction is that provided by the work of Emile Durkheim. This influence is made clear as we discuss the fourth principle of governance. The only thing which needs stressing here about our Durkheimian direction is that we are quite consciously attempting to provoke interest in the strong connection we see between the Durkheimian tradition and the emerging Foucaultian governmentality tradition. Foucault himself is remarkably silent on the possibility of this connection, as are most of his followers and most post-Foucault writers in the Durkheimian tradition, though this latter situation may be changing (Alexander 1988). When one considers Durkheim's pioneering role in promoting the social as a special area of study on the one hand and the Foucaultian interest in 'the invention of the social' on the other, it is hard not to see the connection.

A basic definition of governance

We take governance to be any attempt to control or manage any known object. A 'known object' is an event, a relationship, an animate object, an inanimate object, in fact any phenomenon which human beings try to control or manage. This definition is in some respects circular, in that one of our arguments is that objects of governance are only known through attempts to govern them. We return to this paradox later.

For now, think of any or all of the following: the weather, a romantic relationship, an act of a god, the performance of a company, eating, fighting a war, preventing a war, an apple, a refrigerator. In other words think of any 'thing', 'object', or 'phenomenon'. Now try to think of this thing without the existence of thought (not only your own, any thought at all) about the control or management of that thing, for example, sheltering from the rain, being happy with a romantic partner, appeasing a god, making a profit, eating enough, destroying an enemy, maintaining a peace, apples on trees or in shops, some milk in the refrigerator. We suggest it is very difficult to do so. We return to a discussion of this difficulty at the end of this chapter where we offer an answer to the twin questions: is there anything we can know which is not subject to governance? and can anything be governed completely?

Our definition of governance combines three dictionary definitions: 'government', as in the rule of a nation-state, region, or municipal area; 'self-government', as in control of one's own emotions and behaviour; and 'governor', as in devices fitted to machines to regulate their energy intake and hence control or manage their performance. When we use the term 'governance' we mean the process informing these three aspects, the process of attempting to control or manage a known object. Sometimes we use the term 'governing' as a substitute for 'governance'.

When we use 'government' we are talking about the more particular processes involved in attempting to control or manage a nation-state, region or municipal area. When we discuss attempts to control or manage phenomena which have come to be referred to as aspects of 'the self', or indeed 'the self' *per se,* we use the term 'self-government' or some clearly marked substitute for it, like 'self-control' or 'self-management'. It should be clear that there is an intimate relationship between government and self-government, with the former often attempting to operate via the latter and the latter often taking on the demands of the former as a matter of course. Governing weaves a complex web here.

We do not use the term 'governor' in the somewhat archaic sense given above. We include it as part of our definitional discussion as it is a wonderful metaphor for a crucial aspect of governance, especially law as governance, regulating in order to control or manage performance.

The four principles of governance

In our account governance has four principles. The principles overlap and intertwine to a considerable extent, but we discuss each one in turn for ease of presentation.

Principle 1

> All instances of governance contain elements of attempt and elements of incompleteness (which at times may be seen as failure).

For us, social life is characterised by attempts to control or manage all known objects, including, crucially, other attempts, and by the fact that every attempt falls short of complete control or management. This incompleteness is central to the process of governance whether complete or total control is explicitly attempted. Where only a small amount of control is explicitly sought by the social actor(s) involved, we say the lack of total management is 'incompleteness'. Where an explicit attempt is made to achieve something like complete control over a known object, we label the lack of total management 'failure'.[2]

Consider, first, the governance of unemployment levels by a national government. Government officials may attempt to control unemployment only to the extent of limiting it within current policy considerations; for example, unemployment levels of over 10 per cent have become normal over the last decade, but would have been unacceptable in the 1960s. Even if the government succeeds in this, it must be said to be incomplete governance of unemployment. If it cannot achieve even this and unemployment rises, this instance of governance by government fails. In either case, governance itself continues. Indeed, incompleteness or failure only serves as an incentive for new governing efforts.

Consider, secondly, the governance of a love affair by either party involved. The person concerned may attempt to manage the 'falling-in-love' emotions involved by trying to concentrate on work or going out with friends. Whether the person limits the amount of distraction generated by the affair yet continues the affair (incompleteness), 'falls out of love' by ending the affair yet remains alive and thus eligible for future affairs (incompleteness), or cannot think about anything but the 'in-loveness' which is the known object (failure), governance itself continues. Of course, as Foucault hints in his darker moments, death may provide the only instance of complete governance available (Miller 1993); we explore the implications of this suggestion later.

Consider, thirdly, the governance of a clean bathroom. Whoever attempts to control the cleanliness of this bathroom, they are of course bound to suffer, at best, some dust settling on their room, even if they lock it for a year (incompleteness) or, at worst, someone walking dirty shoes into it just as they finish one cleaning session (failure). In any event, governance continues.

The point of these three examples, as well as illustrating the incompleteness of governance, is to emphasise the perpetual character of governance. Governance does not keep on governing in spite of the incompleteness (or failure) which is so much a part of it, it keeps on governing precisely because and in as much as it involves incompleteness or failure. In this way incompleteness/failure is primary in all aspects of social life. That is, it is always more important than completeness, achievement, or success, despite the fact that many accounts of social life, collective, institutional (by governments, families, companies, churches) and individual (self accounts) focus on completeness, achievement, or success.

Principle 2

> Governance involves power (but only in a very particular sense) and as such involves politics and resistance.

To use a mechanical metaphor, power, for us, is the always-incomplete technical process by which governance drives the machine of society. 'Power' is a technical term involved in the always-incomplete operation of a machine. Just as the term 'power' is commonly used to refer to the technical process by which petrol fuels an incomplete (imperfect) internal combustion engine for it to (imperfectly) drive a car, or the technical process by which coal, water or nuclear fission fuels an incomplete (imperfect) electricity grid to (imperfectly) drive any number of electrical appliances, so, we suggest, we should think of power in society. Engines drive cars and electricity drives appliances 'incompletely' or 'imperfectly' in the sense that they do not operate completely or perfectly, they are not expected to operate perpetually in exactly the same way, something always goes wrong. The only perpetual aspect of the process is the perpetual process of keeping the process going, that is, the perpetual governing. In these examples power is the process of 'keeping things going'; it is not a 'thing', in the way fuel or electricity is.

We make use of the metaphor of society as a machine in introducing power's role in governance because it allows us to highlight important features of this role. However, we do not want the metaphor to get out of hand. Society is not a machine in a simple functional sense. In this sense, the performance of a machine can be assessed against its design. Society is not designed. We discuss the complexities of the notion of society shortly.

So, power is mundanely 'productive' (to use Foucault's term; he sometimes also calls it 'positive' power), it is the technical process whereby all aspects of social life are produced, the process of governance. 'Power' is a summary term for the vast array of governing techniques which come together in various combinations as governance. To be completely clear, in portraying power as the process of governance, we are also portraying it as the techniques which make up this process. In this way, if we so wish, we can summarise the techniques of managing a national economy, statistical, monetary and fiscal, which are used in the governance of the economy as 'national power' or 'state power' (a substitute term for 'state craft'). Similarly, we can summarise the techniques of managing the self in love, emotional, physical, conversational, which are used in the governance of a romance as 'personal power' or 'power over the self' and we can summarise the techniques of managing a clean bathroom, physical, technological, perhaps conversational, which are used in the governance of clean domestic spaces as 'household power' or 'domestic power'. Of course some of these uses of the term 'power' are more common in social and political analysis than others, but they are all consistent with this Foucaultian governance approach to power.

In being mundanely productive, power is definitely not spectacularly 'negative', especially not spectacularly conspiratorial. To say power is the technical process of producing all aspects of social life, the process of governance, and to equate this with the process of producing cars which move or producing electrical appliances which operate is precisely to deny that power is the possession of any individual, group, or organisation which directs the process to its or their own ends (the defining features of what Foucault calls 'negative' power). While we can read the process in terms of the advantage/disadvantage of particular actors at particular times in particular places, as we discuss shortly, to suggest that these outcomes in fact drive the entire process, makes only as much sense as saying that the fact some car engines work better than others drives the process of internal combustion power (the governance of the internal combustion engine), or the fact that some light bulbs are working while others are not drives the process of electric power (the governance of electricity). This brings us neatly to questions of politics and resistance.

'Politics', like 'power', is a summary term. Where 'power' summarises the processes of the operation of various techniques of governance, 'politics' summarises the processes which have emerged and which continue to emerge, in myriad form, concerned with the contestation of power, that is, to be more precise, the contestation of techniques of governance. We take this understanding of politics to be a careful generalisation nurtured by many years of attention to the details of instances of governance (by many scholars of governance, from Machiavelli on), not the type of careless generalisation of 'positions' or 'stances' (as in 'to take up a position' or 'to make a stand') so often associated with the term 'politics'. The politics we are theorising may or may not involve positions or stances; it does not matter, it is their technical relation to governance on which we are focusing.

Politics as the contestation of instances of governance is very much part of the perpetual character of governance. Whatever the known object being subjected to governance, whether it is a national economy, a love affair, a clean bathroom, a disease, a piece of fruit, eating, or a war, one technique of governance is always either being challenged by another technique or awaiting challenge. We may say techniques are challenged for not being the most appropriate technique, but the nature of the challenge – it may be for not being the meanest technique, the kindest technique, the most liberal, the most conservative, the most righteous, most wicked, most efficient or most disruptive – is not as important to our account of governance (though of course it is crucial to any account of any particular instance of governance) as the fact that all techniques are always either being challenged or awaiting challenge. This is the case whether a challenging technique is fully formed, half-baked, or barely embryonic. It is impossible for a technique of governance to be without either challenge or potential challenge; it would not be a technique of governance if it were without either. It must be remembered that governance always involves the cycle 'attempt at control – incompleteness (failure) – attempt at control – incompleteness (failure)', no matter how long the cycle takes.

Consider the governance of a nation's economy and techniques to do with controlling the money supply (monetarist techniques). These techniques are constantly being challenged by other techniques to do with, for instance, controlling demand. The form of the challenge may be that the monetarist techniques are not the most efficient, not the most humane, or not the most accurate in terms of the way the economy works. What is most important for our account is the fact of their being constantly challenged, that is, the fact that they create contests over the governance of the nation's economy and thereby create politics.

To take a less obvious example, consider the self-governance of the 'falling-in-love' emotions of a couple. The persons concerned may feel that they are engaged in no governance at all, that their emotions are swamping them, out of control. This is not possible according to our theory of governance (and recent sociological research supports us; see for example Jackson 1993; Duncombe and Marsden 1993). The parties may well be using this 'let it just happen' technique, but it is a technique nonetheless. At any moment the alternative technique of 'be more sensible, don't let your emotions run away with you' may challenge it, for instance. A contest between techniques is always either happening or about to happen; governance is subject to politics.

We have laid the ground for an account of resistance. Resistance is a technical component of governance, a component heavily involved in the fact that governance is always subject to politics. Resistance is part of the fact that power can only ever make a social machinery run imperfectly or incompletely.

In Foucault's words, resistance is the 'counter-stroke' to power, a metaphor with strong technical, machine-like connotations. Power and resistance are together the governance machine of society, but only in the sense that together they contribute to the truism that 'things never quite work', not in the conspiratorial sense that resistance serves to make power work perfectly.

To reformulate our two previous examples, we can say the challenge presented by demand-management techniques of economic governance to the dominance of monetarist techniques is resistance and we can say that the challenge presented by the 'act sensibly' technique of self-governance to the dominance of the 'let it all go' technique is resistance. We can even say that those individuals, groups and organisations involved in promoting the challenge of demand-management techniques to monetarist techniques are involved in resistance (it would be unusual, though not improper, to describe one 'side' of a person's 'self' as politically resisting another 'side'). However, this should not be taken to mean that resistance *per se* drives governance in some conspiratorial sense. To think about politically inspired groups 'bringing down' a system of governing a national economy through their resistance or to think about 'the system' repressing their resistance is not for us to think about the central feature of politics or to think about the central role of politics in governance/power. Politics and governance operate, as we have described, in a much more technical, usually much more mundane, fashion than this somewhat romantic mode of thought suggests. For us, to see political resistance and the repression of it driving governance makes only as much sense as seeing a car engine being driven by the small cracks in the piston shafts getting together to form a resistance movement and being repressed by the strokes of the pistons.

None of this is to deny that the politics associated with governance sometimes feature 'exploitation' and 'repression'. The 'contests' which are at the heart of our definition of politics can be fierce and bloody, just as they can be passive and mundane. We have concentrated thus far on the passive and mundane aspects of the politics of governance because these are the types of contest which predominate in the world according to our theory of governance. As we are, in this part of the book, promoting a particular type of sociology, it is very important to us that our underlying picture of society (society driven by governance) reflects this predominance of passive and mundane politics. It has to be said that in so doing we are also providing something of a corrective to many years of political theory and analysis which has discussed politics as if it were only about active contestation, especially fierce contestation, often, but not always, using the metaphor of 'struggle'.

We use the somewhat neutral terms 'advantage' and 'disadvantage' in describing the outcomes of the contests which are politics. These terms are appropriate for the full range of outcomes from the passive to the brutal. For us, political analysis is the process of paying attention to the details of a contest at a specific time. This last qualification is crucial; political analysis can only ever be a snapshot of ongoing processes of contestation. While analysis involves attributing the status of advantage to one or more actors and disadvantage to others, this should only ever be taken to be advantage and disadvantage in regard to a particular contest at a particular time.

If analyses of a contest over a long period continue to see one actor or group of actors in a situation of disadvantage, it may be reasonable to understand this long-term disadvantage as structured oppression, and if analyses of a contest over a long period continue to see one actor a group of actors in a situation of advantage, it may be reasonable to understand this long-term advantage as structured exploitation. The instances when it is reasonable to assess technical advantage or disadvantage in the above more emotionally charged terms are instances where the analyst considers that one side has no chance of turning disadvantage into advantage in that particular context. This, of course, is not always a straightforward matter, as anyone will know who tries to attribute advantage/disadvantage to the parties involved in, say, a long-running contest over how to govern profit margins between well-paid airline pilots (employees) and an airline (employer) committed to cheap airfares and battling to stay afloat. Consider, also, the very long-running contest between women employees and all employers in many modern western nations, over how to govern employment relations. While some analysts suggest that women's achievements in securing equal opportunity show that disadvantage can be turned around, others point to the fact that these achievements still apply to only a minority of women employees as evidence of structured disadvantage, and they reasonably call it oppression.

We have no magic solution to this ongoing problem of political analysis. In concentrating on the technical aspects of governance, we are not trying to side-step the difficulties of political analysis, merely trying to clarify them and provide new tools for tackling them.

One final point about power, politics and resistance concerns the darker side of resistance we hinted at earlier when we mentioned Foucault's attitude to death; here our ground is not so well laid. James Miller's biography of Foucault (1993), especially, provides much food for thought on this darker side for those, like us, trying to build a Foucaultian approach to the study of social life. By 'darker side' we understand something like an imperative to resist. Clearly, we are here on the terrain sociologists have traditionally called 'the irrational'. We attempt to map this terrain, or at least acknowledge its role, more inspired by Durkheim's commitment to include it in his basic account of society than by Foucault's determination to celebrate it; Miller's account makes 'determination' an appropriate word.

What does an imperative to resist look like? This is a difficult question to answer sociologically as so little sociology has been prepared to examine it. We might direct the reader to the philosophy of Nietzsche or even the more literary writings of Georges Bataille and leave it at that. While this would be helpful, it would be to refuse the sociological challenge involved. We take up this challenge in offering the following brief account of what we understand Foucault to mean, bolstered by a summary of Jack Katz's fascinating *Seductions of Crime* (1988).

The imperative to resist is an urge to transgress, an urge to move to another level of contestation. In this sense, resistance is the dark side of politics. As we said earlier, contestation is a necessary part of governance. Our account so far makes this seem a fairly rational process; rational techniques for governing the economy being challenged by other techniques, rational techniques for governing the 'self in love' being challenged by other techniques, the same is true for the clean bathroom. In each of our examples the alternative techniques appear just as rational as those they are challenging. Perhaps in the 'self in love' example we are giving a slight flavour of Foucault's darker side of resistance. We can develop this example and lead it into a discussion of both Foucault's and Katz's interest in the matter to begin to capture the irrational element involved (probably 'extra-rational' is a more accurate term, but we stick to 'irrational' in an attempt to remain close to standard sociological ground).

Consider a situation in which a man in love, who is experiencing an inner contest, 'inner politics' if you like, along the lines set out previously, 'Be sensible, control yourself, don't let your emotions get the better of you' versus 'Let go, let your in-love passions have their head', suddenly sees the woman who is the object of his in-loveness in an affectionate embrace with another man. A third, dark side to the contest of governance suddenly appears, a new and shocking technique of governance is available: 'Lash out at her, hurt her', perhaps intermingling with 'Lash out at them both, hurt them'. The picture is instantly less rational. Our man may do nothing (a victory for the 'control yourself' side), he may try to hurt by telling the woman he's no longer interested in her, he may attempt to hurt by a spray of insults, he may strike one or both parties, or, in extreme instances, he may kill. In any case, what is important for our account of governance is that an imperative to resist the two techniques in contest is produced by a new, dark technique, which for at least an instant appears the superior technique, a sort of rationality of irrationality dramatically takes over the scene of governance.

Foucault's work on the case of Pierre Rivière (*I, Pierre Rivière* 1975) is an indication of his long-held interest in the imperative to resist. Katz (1988), inspired more by ethnomethodology and symbolic interactionism than by Foucault, provides a great deal of sociological data on those moments when the imperative to resist leads to criminal behaviour. He pays attention to the detail of passion killing, shop-lifting, 'gang' violence, robbery and 'cold-blooded' murder. By concentrating on the sociological foreground of these instances of governance (rather than on background factors like class, race and gender, as so much sociology does), Katz amply demonstrates the force of this imperative in these instances, or 'seductions', as he aptly calls them (his book is subtitled 'Moral and Sensual Attractions in Doing Evil').

Foucault suggests, according to Miller's well-documented account, that the imperative to resist is a search for freedom from governance and as we have seen he hints that only death can bring such freedom (Miller 1993). He also suggests that this imperative drove Foucault's own transgressive behaviour; hence, his keenness to live out some of the detail of the Marquis de Sade's bizarre searches for pleasure.

Are we straying too far from a balanced account of governance here? The reader might well suggest that in dealing with the imperative to resist, we are dealing only with the governance of the self and only then with that tiny fraction of instances where the politics of the governance of the self spills spectacularly into dramatic transgressions. We now complete our account of the role in governance of the imperative to resist by broadening our description to the more mundane aspects of governance at the centre of our portrayal.

If we move from Foucault and Katz to F.G. Bailey's stimulating *The Tactical Uses of Passion* (1983), we find some data on much less dramatic but nonetheless effective interventions by the imperative to resist, this time in cases of university governance involving committees. Here we see the imperative at work in an organisational setting, not tempting individuals to violence, but taking committees in directions that no single member or group of members could have predicted in advance. This, we suggest, is a fairly common experience in the governance of organisations. A room full of people of seemingly unified intent, or at least similar intent, a union meeting perhaps, finding themselves subject to tensions and divisions, sometimes subtle and temporary, sometimes not, which no one of them would wholly own yet all of them would collectively recognise as influential on the outcome of their meeting.

In short, we are drawing attention to the mysterious aspect of the incompleteness of governance which is so commonly experienced, often in a 'devilish' guise, whether excitingly or annoyingly so, rather than in an evil guise, when 'things never quite work': the children misbehaving, the car window being stiff, a romantic liaison not going well, an unexpected bill arriving, a work meeting producing trivial annoying tasks. In another register this also includes such things as: a group turning violent in a pub because someone 'looked at them', a person shoplifting 'just because the goods were there', and a national government taking offence at remarks made by a leader from another nation and instigating military action against the 'offending' nation.

Principle 3

Governance always involves knowledge.

Our theory of governance sees two crucial roles for knowledge. Knowledge is used to select objects for governance and knowledge is used in the actual instances of governance.

In dealing with the first of the two roles, we immediately confront the paradox that we highlighted at the beginning of this chapter: while knowledge is used to select objects for governance, the objects of governance are only ever known through governance. We employ two theoretical tactics in overcoming the obstacles thrown up by this paradox: the notion of the 'always-already' and the assertion of the primacy of governance.

Following Althusser and Foucault, we refuse the necessity of origins in theorising governance. While our formulations involve objects which have histories, our concern with the past of these objects is genealogical, that is, we follow Foucault's method which uses accounts of the past to disturb the present, to 'render it strange', rather than exploring the past to discover the origins of the present (Bevis et al 1989). In this way (and this is Althusser's formulation), the objects we discuss as objects of governance are 'always-already' there, they have a past, they may even have beginnings, but they do not have origins in the sense of a genesis which completely determines their form (Althusser 1969).

When we speak of an economy being governed, we do not propose that to govern it, say to bring inflation down, it is necessary to know its origins. While many facts about the past of this economy may be known, governance always treats it as always-already there. The origins of the economy are not important. Inflation can be checked without knowledge of the origins of the economy from beyond this governance of it. As we stress, governance is a cycle, 'attempt at control – incompleteness (failure) – attempt at control'; it is a cycle without end and a cycle without origins, for itself and for its objects.

To support this tactic against the effects of 'the paradox of the known objects of governance' we assert that governance is always more important in social life than the known objects governed and hence than the knowledge of objects. This assertion may appear something like an assertion that the egg comes before the chicken (or *vice versa*), but its place in our theory of governance should not be underestimated. Just as we argued earlier that incompleteness/failure is always more important than completeness, achievement or success, despite the fact that completeness/success is usually treated as central, so for us here, governing a known object is always more important than the knowledge of the object, despite the usually assumed centrality of knowledge.

When inflation in an economy is subject to governance, or a clean bathroom, or a romantic relationship, or eating, or any known object, knowledge of the object does not exist independently of the governing, before the governing comes along to do its governing; knowledge does not force the hand of governance. Rather, we assert that the reverse is the case.

Governance exists independently of knowledge; it leads to the knowledge of objects being governed precisely in governing them. Inflation is only a known object in that it has been and continues to be subjected to governance. We may eat, love, or fight before we know about it (while this seems a concession to intuitionism, we allow it to show that even it does not disturb our argument), but we do not know about it before we attempt to control or manage it, that is, before we govern it. Of course, this is easier to grasp with more abstract objects, like inflation, than it is with these 'natural' objects.

Returning to our first role for knowledge in governance, knowledge used to select objects, the above discussion may be thought to render this role marginal. If knowledge is used to select objects for governance, yet knowledge is led by governance, are we not saying that governance is used to select objects for governance, thereby pushing knowledge aside? Our answer is 'yes' but 'no'. Yes, in one sense, governance is used to select objects for governance, but no, this does not push knowledge aside.

Just as the process whereby the desire for a clean floor leads to the floor being cleaned and hence to a clean floor does not render the process of cleaning trivial, does not lead us seriously to believe that the desire for a clean floor cleans the floor, so it is with governance and knowledge. Knowledge actually does the selecting work for governance.

This role for knowledge in governance involves only very basic governance, what we call the primary level of governance. This is the 'always-already' level. All objects of governance, as we discussed above, are always-already objects of governance. Knowledge works for governance in a seamless, invisible process. Knowledge selects objects for governance by posing and answering questions like 'what is an economy?' (making an economy subject to governance), 'what is a war?' (making a war subject to governance), 'who am I?' and 'what are these feelings I am having?' (making a self-in-love subject to governance). The objects involved are governed in that the basic acknowledgement of existence which knowledge performs involves a basic attempt at control or management.

In line with this primary role for knowledge in first-order governance, we could not know objects if they were not always-already governed, we could not be writing about them if they were not. We need say no more about this primary role for knowledge; it is crucial to governance, yet its seamlessness means we cannot say more than we have without turning to the intricacies of metaphysics, which are well beyond the scope of this book.

It is on the secondary level of governance involved in the second (and secondary) role we see for knowledge in governance; knowledge is used in the actual instances of governance and thus the knowledge process is accessible to sociology. On this level knowledge is used in choosing and implementing techniques of governance beyond the basic 'acknowledgement of existence' technique. This is where sociology must begin its interest in governance. As we argue in our methodology chapter

(Chapter 6), this is where sociology must begin to ask 'how?' (not 'why?').

Knowledge is involved in a different way in this second level in the basic attempt at control or management. Here an acknowledgement of existence is supplemented by attempts to impose more control or management: the economy should be slowed, the economy should be stimulated; the war should be stepped up, the war should be ended; 'I must control my in-love urges', 'I must see my in-loved one'; 'the bathroom should be cleaned today', 'the bathroom will be fine until next week'. Here knowledge is being used to select some techniques of governance over others and to implement the chosen techniques in the attempts to impose control or management on the objects concerned.

The knowledge used ranges from very simple, informal knowledge to very complex, formal knowledge and the range includes knowledge called rational, within modern social sciences, and knowledge called irrational knowledge. Each technique of governance may include some simple, informal (perhaps irrational) knowledge and some complex, formal (very rational knowledge).

Consider, as a first example, the 'control your in-love emotions' technique for governing the self in love. This technique is informed by a combination of: simple, informal knowledge to do with taking care of the self passed on in 'commonsense' conversations between friends or between parents and children; 'irrational' knowledge to do with religion (the necessity to control one's emotions so as not to offend a god or gods) or superstition (the necessity not to think too much about the object of one's in-loveness for fear of 'putting them off'); and complex, formal (very rational) knowledge to do with psychological, medical, or sociological theories and experiments concerned to determine scientifically the effects of emotional control on psychological, medical, or social well-being passed on through formal practitioners' advice, newspaper and magazine articles, or formal educational mechanisms.

Even the governance of an economy, to take a second example, involves techniques which combine knowledge in a similar way. Monetarist techniques for governing inflation, for instance, are informed by a combination of: complex, formal (very rational) knowledge based on economic theories and models designed to determine the effects of changes in the money supply on economic well-being which is passed on in formal government documents, journal articles and newspapers; simple, informal knowledge to do with which policies are likely to find favour with government officials or international bankers which is passed on in informal conversations in hallways and tea-rooms; and, in addition, not a little blind faith that this is 'how people really are' (sometimes called voodoo economics).

This is not to rule out the possibility that some techniques of governance are informed by purely formal knowledge and some by purely informal, though we contend that the use of such unitary techniques is very rare, in the modern, western world at least.

Foucault's work is, of course, central to our thinking about this second role for knowledge in governance. Three closely connected points need to be stressed in the light of Foucault's direct influence. First, the knowledge which is used in the actual governing of objects is always available knowledge. We stress 'available' because Foucault elaborates the ways some knowledge is made available by the operations of the institutions involved in instances of governance while other knowledge is not made available. It is in elaborating the complex relations between definite institutions and available knowledge that Foucault's generalisations about the connections between governance and knowledge are best seen.

Second, Foucault's work details the rise of those formal knowledge complexes known as the human sciences and traces many of the ways in which they have come to inform widely used techniques of governance in the modern world: new knowledge of madness generated by psychiatry being used in the governance of deviance; new knowledge of punishment generated by penology and psychology being used in the governance of nations and regions.

Third, Foucault indicates, an indication taken up strongly by some of his followers (Hacking 1975, 1990, 1991; Rose 1991; Miller and Rose 1990), the importance of statistical knowledge for many modern techniques of governance. This point is especially relevant to the government of modern nation-states, but it is not without its importance for consideration of governance more generally, as we can show by returning to one of our two previous examples of the knowledge-governance nexus.

The knowledge which informs the 'control your in-love emotions' technique for governing the self in love is made available by the operations of institutions to do with self-control (education, counselling, newspapers, magazines) and in-loveness (magazines, novels, poetry, films and television programmes). In the late twentieth-century western world these institutions make this knowledge generally available at the expense of other knowledge, like sorcery, which is not generally now available. As we hinted above, this technique is also informed by the formal human sciences, especially psychology and sexology, but also biology. This formal knowledge informs this technique even where it appears to be a 'private' usage (a person 'privately in love'); such is the reach of the modern human sciences that there is no private space for thinking about the self beyond their reach (Rose 1990, 1992). It is in this vein that we can insist that statistics reach even techniques of governance far removed from the government of modern nation-states or regions. The 'control your in-love emotions' technique, informed as it is by the human sciences, is indirectly informed by statistics which both use and are used by the human sciences in maintaining their importance for governance. While only some people may undertake statistical calculations about the chances of their attempt to control their own in-loveness, all thinking about controlling the in-love emotions in the modern world must be informed by some aspects of modern psychology, sexology, or biology, however indirectly, which all rely heavily on statistics in going about their business.

Principle 4

> Governance is always social and always works to bind societies together (which sometimes, ironically, involves social division).

In elaborating this principle, we need first to define the terms 'social' and 'society' (terms which we have used freely thus far). We use both terms in two related but distinct senses. In each sense we use the two terms interchangeably, in the standard manner; that is, we use 'social' as an adjectival form of 'society' (we also use 'the social' as an alternative, if somewhat more particular, noun form of 'society'). We refer to one of our distinct senses of 'society' and 'social' as traditional, the other as Foucaultian.

The traditional sense of 'society' and 'social' is that which pre-exists individuals in consideration of the collective actions of individuals. This sense goes back to at least the ancient Greeks, but for sociology it is most clearly and extremely stated by Durkheim. We say 'extremely' stated because Durkheim uses 'society' and 'social' as that which pre-exists individuals in consideration of the collective and individual actions of individuals. Despite evidence of some qualification by Durkheim on this extreme proposition (Nisbet 1965: 49–53), we stick to his extreme form when using the traditional sense of 'society' and 'social'.

In reading Durkheim in this way we make yet another return to Althusser's notion of the always-already. For us, society (and the social) is that which always-already pre-exists individuals in consideration of their collective and individual actions. In this way, society is always-already there. There is no point searching for the origins of society, in this traditional sense, for all one will ever find is society. Sociology is based on this productive tautology: society is always-already there and the always-already is society. Sociology is the study of how society is always-already there; that is, to labour the point, sociology is the study of how – the exact detail of how – society is society.

The Foucaultian sense of 'society' and 'social', most often presented by its adherents in the alternative noun form 'the social', stems from some of Foucault's work around the notion of governmentality (Foucault 1979) and from some work by some of his followers, especially Donzelot (Donzelot 1988, 1991). This sense of society or the social focuses on its 'invention' as a definite category of the government of nation-states and regions in the nineteenth century and on the development of this invention in the twentieth century. In this Foucaultian sense, society is a modern, western phenomenon.

Society or the social is a new conjunction of certain concerns about populations, their health, their longevity, their education, their children. These concerns are the targets for new governmental techniques which we might nowadays summarise in English by the term 'welfare'. The stimulus for the invention and development of this sense of society was and continues to be some advances in insurance technology. In line with some developments in probability theory it became possible to think about, to make definite calculations about, to make policy about, and to make provision for, the health and longevity of larger and larger numbers of people. While, obviously, governments were able to think about these concerns before, even to make policy about them, the devastating new ingredients were the fine detail of the calculations and, crucially, a device whereby governments could attempt to provide financially for the future of the population on the basis of these calculations (Hacking 1975; Ewald 1991; Defert 1991).

This Foucaultian sense of society dovetails with the traditional sense. The fact that society is always-already there is boosted by the emergence of a new field of government around 'the social'. From the late nineteenth century 'society' attracted an enormous amount of attention, which in turn encouraged much more thinking, talking and writing about its always-alreadyness. This situation continues as we approach the end of the twentieth century. While sociology, as we say, is based on the productive tautology of society as always-already, it owes its emergence and continuation as a social science to the 'invention' of the social.

All instances of governance are social (part of society) in the traditional sense though not in the Foucaultian sense. It will be remembered that the objects of governance are always-already known through governance. In this way governance is always social. In every instance of governance the object of governance and the techniques of governance are made available by society, they are always-already available. Whether it is a self in love being governed or ethnic strife in eastern Europe or any other object, this is the case.

To work briefly though these two concrete examples, both the self in love and ethnic strife are always-already available to the actors involved, they are socially available objects, they have no existence beyond society. Whatever techniques of governance are employed in these two instances, quiet contemplation to control the emotions, conversation with friends, international sanctions against states, warfare, United Nations negotiations or intervention, they are always-already available, they are socially available techniques, they have no existence beyond society.

For the Foucaultian sense of society, of course, only some objects of governance are available – those invented as part of the invention of the social, like children's welfare, infant health, and mass literacy – and only some techniques of governance are made available – those invented as part of the invention of the social, like child protection legislation, infant health programmes, and mass schooling (these techniques were invented in just the same way as inoculation was invented to control contagious diseases). We can sensibly say, then, that this principle of governance is more Durkheimian than Foucaultian. It is in the light of this proposition that we discuss the second half of this principle of governance: that governance is always working to bind societies together. Our account draws especially on Alexander (1988), Collins (1985), Durkheim (1965) and Nisbet (1965).

We make clear at the outset that we do not subscribe to a conservative functionalist reading of Durkheim. That is, we do not propose that societies are always-already organised in such a way as to promote social unity. Rather, as we hinted in our wording of Principle 4 and as we discuss in more detail later, we believe the binding process involves separation and division. Binding, for us, in being part of governance, is never complete, it is always provisional and unstable.

While society is always-already there, this does not mean, of itself, that it is always-already there strongly enough to bind actors into it. Being always-already there might entail, for example, no more than the pre-existence of in-loveness and self-identity for the governance of the self in love or the pre-existence of agonistic ethnicities and national identities for the governance of ethnic strife, for example, in eastern Europe; it does not necessarily entail actors being bound to one another around these pre-existing objects. For actors to be bound together, whether individuals or organisations, governance has to work to strengthen the always-alreadyness of society, to add this extra dimension of boundedness to the traditional meaning of 'society' and 'social'.

It must be remembered at this point that governance is perpetual (as, of course, is society) in being always incomplete or failed. So, it must be realised that governance's work in binding societies together is always incomplete or failed. Governance in this guise attempts to recreate social solidarity as each attempt fails or is incomplete. This incompleteness/failure results from the crucial role played by division and exclusion in those very processes by which the binding is attempted. Moral and ethical discourses, as we discuss in detail shortly, construct 'the bad' precisely in discovering and valorising 'the good'. This leads to the constitution of the separated or excluded 'other'. It is in this sense that Durkheim speaks of societies needing crime (Durkheim 1964: 72); the privileged norms are reinforced by the reaction against the transgressor. Similarly, Foucault speaks of the importance of dividing practices which set apart the insane and the sane, the undeserving poor and the deserving poor (S&P 1982).

Morality, community, communications, physical structures and sacred rituals (whether formally religious or not) are mechanisms for this type of binding social governance pointed to by Durkheim. Morality is a mechanism of binding social governance in that it attempts to unify individuals and organisations (churches, companies, schools) around particular themes of right and wrong. For this account the content of the morality does not matter. The attempts at binding may feature Christian morality, pagan morality, criminal morality, or whatever. What is important is that individuals and organisations are bound together by them being guided/coerced/encouraged/induced towards certain ways of doing right and away from certain ways of doing wrong. What is also important is that the binding never works completely; at least some individuals and organisations slip through the nets of morality, for longer or shorter periods.

'Community' is a summary term here for certain techniques of intimacy, continuity and cohesion. The mechanism of community operates alongside morality in attempting to unify individuals and organisations around particular themes of right and wrong. Again, the content is not what is important, it is the incomplete binding. Community contributes certain themes to the right/wrong complex, like themes of place or ethnicity. Hence, local communities or ethnic communities are particular forms of the community mechanism going about its (always incomplete) binding work. This mechanism constructs 'others' as different and potentially hostile.

Communications and physical structures also work together in this Durkheimian picture of social binding. They do their work in different ways in different societies, depending largely on the number and geographical spread of individuals and/or organisations involved. They contribute to the morality-community mix in these different ways. In societies with relatively small numbers of individuals and organisations (possibly with no or very few organisations, as we have come to know them in the modern west) spread over a relatively small area, communications devices include immediate verbal communication and limited forms of stored communication and communication over distance (which may or may not involve forms of writing). Physical structures are similarly simple, though religious factors may mean quite complex physical structures are built. In these types of societies, sometimes referred to as primitive or simple, communications and physical structures, while their binding work is always incomplete of course, have less binding work to do than in societies with relatively greater numbers of individuals and organisations and/or with relatively greater distances between them.

In these more complex societies communications devices include a much greater array of forms of stored communication and communication over distance, as well as immediate verbal communications. Some form of writing is always involved; the nineteenth and twentieth centuries have seen the development of very complex electronic communications devices used for both transmission and storage. Physical structures are similarly complex as these large numbers of individuals and organisations give physical form to their housing, working, recreating and worshipping activities. For the Durkheimian account, networks of modern cities are the most complex mixes of communications and physical structures ever developed. It is no surprise, then, that this account strongly suggests that the binding work of these mechanisms in these types of societies is so demanding that the inevitable incompleteness is much more apparent than it is with less complex societies.

Straddling the four mechanisms of social binding discussed so far and, in an important sense, informing them all is the mechanism of the sacred. This mechanism operates by making some objects or practices sacred and others profane. The binding is attempted both around the sacred and against the profane. Formal religions are the most obvious concrete example of this mechanism, working hard as they do to make some objects (churches, totems, statues, bibles) and some practices (prayer, confession, sacrifice) sacred and other objects (consumer goods, pornography) and other practices (gambling, non-procreative sex) profane. However, in discussing such obvious examples, two important aspects of the sacred may be missed. First, even in cases of formal religion the content of the sacred and the profane is not fixed; for the purpose of binding social governance it would not matter if, for instance, the above lists were reversed and a religion made pornography and gambling sacred and churches and prayer profane; what matters is that the attribution of sacredness and profanity binds individuals and organisations into a society through mechanisms of unification and separation. Second, the sacred operates across all spheres of social life, not just formal religion; sports consumption, for instance, involves making some objects (teams, flags, bodies) and some practices (attendance at matches, keeping score) sacred and other objects (other teams, other sports) and some practices (missing matches, being uninterested) profane.

Governance works to bind societies together (remembering, of course, that it never does so totally or successfully) even in cases which appear at first glance to be dysfunctional for binding, where attempts are made to limit or displace binding mechanisms (anti-religious dissent, political revolution). We contend, along Durkheimian lines, and this, we suggest, is very similar to the Foucaultian point, already made, that power always involves resistance, that there can be no genuine anti-governance activities short of death. There can be only anti-particular-technique(s)-of-governance activities and therefore no genuine anti-binding activities short of death. What appear anti-binding activities may be anti-particular-binding-techniques (like religion) but they must involve some alternative binding techniques. For example, binding may be attempted around wickedness rather than godliness, around revolution rather than stability, or, as we hinted in discussing Katz's work, around criminality rather than obedience to the law. In all cases the mechanism of the sacred still does its work, as do the other four mechanisms we have outlined.

This Durkheimian line of argument often raises the charge of 'functionalism'. As we said, we regard this charge in our case (and Durkheim's and Foucault's) as misleading. In arguing that there is no escape from governance (except perhaps in death and certainly in some extreme instances we discuss shortly in the conclusion to this chapter) we are not arguing that governance is complete. Quite the reverse. We are not therefore arguing that governance makes societies function perfectly. If there is a functionalism involved here, it is no more than the unremarkable claim that societies perpetually reproduce themselves in some form or another. Perhaps our argument that the incompleteness/failure of governance drives this perpetual reproducing is remarkable, but it is more dysfunctionalism than it is functionalism.

Conclusion

We have defined governance and we have elaborated the four principles of governance in line with our reading of Foucault's work and in line with a reading of some of Durkheim's work. We are now ready to turn to law as a form of governance, to outline the content concerns for our new sociology of law as governance. One loose end remains to be tied up in this chapter, concerning the extremes of governance. As promised, we approach the extremes of governance by answering the twin questions, is there anything we can know which is not subject to governance? and, can anything be governed completely?

The answer is yes, but only if we answer the two questions simultaneously. In other words, the only thing we can know which is not governed is that which is governed completely. This is not as convoluted as it may first appear. It suggests a circle of governance. The circle is not visible as a circle from any point on the circle and it is not possible to be anywhere but at a point on the circle. However, a hint about the circle is given, but only at the point where the circle joins. This is easier to grasp through an example. This is the only example we are able to think of and neither of us makes a claim to be able to do more than think about it. As far as we understand, the ultimate Zen Buddhist experience is to achieve a sort of nothingness (David-Neel 1977; Humphreys 1949), a thingness which is not governed. Yet the only way to achieve this act is through a disciplined practice so intense it equates to what we regard as complete governance. That which is completely governed meets that which is not governed.

Of course, another possible example of a thing which is not governed is death, as we have suggested several times in acknowledging Foucault's hints. This does not fall within the scope of our twin questions as it is not something we can know. Perhaps, as some branches of Buddhism propose, death can be completely governed such that it is an escape from governance for those who actively achieve a certain type of death (Rinpoche 1992; Whitton and Fisher 1986), but this possibility is not explored by Foucault and is outside the scope of this book.

5
Law as Governance

Introduction

Our new sociology of law as governance follows directly the four principles of governance detailed in the previous chapter. In this chapter we show that all operations of law are instances of governance by discussing law in terms of each of these four principles. We thereby assert that all operations of law are distinctive instances of governance. What is an 'operation of law'? Needless to say, we are not engaged here in the time-honoured jurisprudential exercise of deciding 'what is law'? We need a basic sociological understanding of law in place.

We are working with a modified version of Weber's famous sociological definition of law:

> An order will be called law if it is externally guaranteed by the probability that coercion (physical or psychological), to bring about conformity or avenge violation, will be applied by a staff of people holding themselves specially ready for that purpose. (Weber 1954: 5)

We modify and clarify this definition in five related ways. The first concerns the notion of an order. Here we substitute the word 'operation'. In this way we stress that law is doing. The slogan for our sociology of law as governance is 'the law is what the law does'. Our second modification concerns probability as an external guarantee. We understand by this, following closely our understanding of Foucault's governmentality, that a calculation is made about probability by some actor or other (whether individual or collective) using a definite means of calculation. This is a crucial aspect of our modified Weberian definition of law, as we show shortly. Our other three modifications are more points of clarification.

Third, we read Weber's physical or psychological coercion very broadly, such that it includes any tendency towards conformity with an historically received norm or the avengement of a violation of an historically received norm involved in the calculation(s) being made. Fourth, following Weber (and here Weber and Durkheim are very much on the same ground), we understand a 'staff' to include any specifically nominated person or persons (in its broadest legal sense, covering organisations as well as individuals), not just modern bureaucratically organised mechanisms. For Weber, of course, it is the specialisation involved, the 'holding themselves specially ready', whether it is the clan acting in a blood feud or the judge sitting in a modern courtroom, which separates law from custom.

Finally (and we have each elsewhere, separately, made this point, albeit in different contexts; see Hunt 1992, Wickham 1987) we understand 'law' as a convenient shorthand term for the diverse operations of diverse laws. In other words, any time we use 'the law' or 'law' (singular) we are using it in this way, as a shorthand. We are especially keen to avoid the tendency to allow the definite article to signal law as an essence of society; 'the law' is thereby taken to be the necessary component, or even a necessary component of a society, something which is essential to its existence, which defines and organises its existence, in much the same way that crude Marxist accounts of a society posit 'the economy' as essential or crude liberal accounts posit 'the individual' as essential. For us, 'law' and 'the law', while used in singular form as a shorthand, indicate very plural entities, namely, the varied operations of different laws. Speaking of 'law' or 'the law' in referring to these diverse operations should be read as no more essentialist than speaking of 'procedure' or 'the procedure' when referring to the diverse operations of bureaucracies. Just as one can speak of bureaucracies following 'procedure' or even 'the procedure' when referring to the vast array of bureaucratic procedures in operation in any modern western nation, without invoking the idea that modern western societies are essentially about 'the procedure', so we speak of 'law' or 'the law' without invoking the idea of the essentialism of law.

So, our working sociological definition of law (using Weber, modified and clarified by our Foucaultian/Durkheimian concern with governance) looks like this:

> An operation is called law where it involves a calculation by some actor or other (individual or collective), using a definite means of calculation, towards conformity with an historically received norm or the avengement of a violation of such a norm, where a staff holding themselves specially ready for directing the conformity and/or conducting the avengement is involved.

We conclude this introduction with concrete clarification of this abstract definition by means of two examples.

All steps involved in the prosecution of a thief (the theft itself, the investigation, the conviction, the punishment) can be said to be operations of law in terms of our working definition. For a theft to be reported (even if only to a friend or family member) some calculation must be made that a violation of an historically received norm about property has occurred. The calculation is definitely made by the victim and probably by the thief; for our definition it is the fact that a calculation is made that is important, not who or which body makes it (though of course this is important for a full analysis of any concrete instance). In the case of theft, no calculation means no theft (which holds for all operations of law; no calculation towards conformity or avengement, no operation of law). The means of calculation involved may be informal, like the victim deciding how much loss he or she has suffered and/or the thief deciding how much he or she has gained and/or both deciding how wrong the theft is and/or the thief deciding how pleasurable the theft is. The means of calculation may also be formal, like an insurance company calculating the loss involved and/or the police deciding whether a crime has been committed (the police in most countries decide in favour of a formal charge only where property of a certain value is acquired by certain means; for example police rarely consider 'taking-by-finding' money on the street as theft). Of course, at some point in the process a calculation must involve a specially ready staff for directing conformity or conducting avengement if we are to understand the theft as an operation of law. Even if it is only that the thief or the victim makes some calculation about any possible involvement whatsoever by a police force, or a court, or a prison system, or a blood feud dispenser of justice, or a church dispenser of justice, we can call the theft an operation of law.

We stress again that the sociology of law as governance needs evidence of a definite calculation using a definite means of calculation, whether as informal as thought or discussion based on an informal training in prevailing morality, or as formal as a legal hearing, before it can recognise an object of inquiry.

This discussion makes it much easier to recognise the other steps in our theft example as aspects of the operation of law. In each of these steps, investigation, conviction, punishment, recognisable calculations and definite means of calculation are involved and at least one actor is involved as a special staff. Investigating a theft may or may not involve a 'specially ready' force of investigation, like a police force, or church inquiry, or tribal inquiry, but it must involve calculations, using a definite means of calculation to do with establishing responsibility or guilt. The prospect of a special event around conviction and/or punishment means the presence of at least one specially ready staff actor, no matter how temporary, in the investigative calculations.

In those instances where formal conviction is involved the specially ready staff are most apparent. A court trial, an inquisitorial trial, a tribal ceremony around the responsible party, all demonstrate the operation of definite calculation mechanisms by specially ready staff in being obvious operations of law. This is also the case for instances involving formal punishment. Formal conviction sets off formal punishment whereby specially ready staff make calculations, using definite means of calculation, whether to do with vengeance, or rehabilitation, or restoration, or all three, about how to imprison, or execute, or corporally punish.

Our second example is more mundane, focusing more on the 'directing the conformity' aspect of our definition of law. Those procedures of a company concerned with proper bookkeeping can sensibly be termed operations of law, even though they are usually less dramatic than the operations of law which make up our theft example. Bookkeeping involves daily calculations by company officials, using a quite formal means of accounting calculation, towards conformity with an historically received norm of corporate accountability. This process is always undertaken with regard to (even if the regard is concerned to avoid) particular legislation to do with proper reporting of company performance policed by a variety of officials holding themselves ready for the purpose, tax auditors, securities regulators.

In this example the pervasiveness of law in the modern era is easy to see. For us, this is a good example precisely because it is so humdrum. Law is operating here with all its key features – calculations, definite means of calculation, historically received norms, specially ready staff for directing conformity and/or conducting avengement – but without fanfare. Law is operating as part of the routine practice of daily life, not as a separate guide to it. Our sociology of law as governance is interested in all operations of law as instances of governance; it must thrive on the mundane as well as account for the spectacular.

The four principles of law as governance

Principle 1

> All instances of law as governance contain elements of attempt and elements of incompleteness (which at times may be seen as failure).

Operations of law contain elements of attempt in a very obvious sense. In line with our sociological definition of law, every instance of law as governance contains elements of an attempt to achieve conformity with an historically received norm or an attempt to achieve the avengement of a violation of an historically received norm.

Law is incomplete just as much as any non-law instance of governance is incomplete. Law fails sometimes, just as non-law instances of governance do. Consider, first, the governance via contract of a business relationship between a component supplier and a manufacturer (Macaulay 1963). This is an instance of law as governance.

However, the contract never governs the business relationship completely. As Macaulay demonstrates, many aspects of the relationship escape governance by contract law, usually and routinely because the parties to the contract ignore the provisions of the contract and go about their relationship in ways that make sense to them as regular actors in this environment but which do not fit the logic of contract law. This is incompleteness. Of course, as we hinted above, sometimes one party or the other explicitly seeks the sort of total control of the environment which the logic of contract law fosters and when this is not attained takes action against the other party for breach of contract. On these occasions incompleteness becomes failure.

Consider, secondly, the governance of a passion killing. We saw in the last chapter that governance is involved in a case like this in terms of self-government and in terms of the imperative to resist governance. In thinking the example in terms of the incompleteness/failure of law as governance we must add several factors. Katz presents convincing evidence that passion killers rarely make calculations about the consequences of their action in terms of formal legal mechanisms like police, courts and prisons (calculating instead in terms of the righteousness of their deeds) (1988: Ch 1). We may wonder, then, whether this is an instance of law as governance at all (remember, no calculation towards conformity or avengement, no operation of law). But Katz's evidence also points out that by the same token of righteousness, passion killers rarely attempt to hide their action either. This means of course that formal legal mechanisms like police, courts and prisons come into the picture anyway, accompanied by their vast array of calculation mechanisms and their many calculations. The passion killing is definitely an instance of law as governance; we would not know it otherwise.

The very fact that the killer usually neither worries about the formal legal mechanisms at the time of the killing nor attempts to elude them immediately afterwards is itself evidence of the incompleteness, even failure, of law as governance in this case. If we regard the objective of law as governance in cases of passion killings as the prevention of such killings, any passion killing indicates the failure of law as governance. Even if we regard the objective of law as governance in cases of passion killings as vengeance against the killers, any case in which the killer regards the punishment as a reasonable price to pay for the righteous act they have committed (and Katz indicates that there are many such cases) means at least that law as the governance of passion killings is incomplete.

All instances of law as governance contain elements of incompleteness/failure in the sense that the law is always chasing at least one objective it cannot catch.

Principle 2

> Law as governance involves power and as such involves politics and resistance.

As power for us is a technical process by which governance drives the machine of society, law, for us, is part of this technical process. Law is a part of power; it helps drive society incompletely or imperfectly. Law is a part of mundanely productive power, helping power produce all aspects of social life. Law is neatly tied into the equation whereby 'power' is simply another term for the process of governance. Just as we can summarise certain governmental techniques as 'state power', so we can summarise legal techniques of governance (legal rules, court procedures, police procedures) as 'law power' or 'legal power', if we so wish. We do not urge the use of such a term but we make this point anyway in an attempt to ensure the reader cannot understand law as separate from and in the service of some mysterious removed power. Law is power in the very productive way that law is governance.

In being mundanely productive, law is definitely not spectacularly negative, especially not spectacularly conspiratorial. Law is not the possession of any individual, group, or organisation which directs legal processes to its or their own ends. 'Law', as a part of power/governance, refers to the technical processes which produce all operations of law, as per our definition.

It should be noted that law has become mundanely productive as part of a historical process which has seen the reduction of its participation in the spectacle of power. Historically many forms of power, military, political and legal, were occasional but spectacular in their operation. They could descend with sudden and terrifying force, but they lacked the capacity for sustained and extensive exercise. As more extensive mechanisms of power became available, for example by expansion of state agencies into dispersed localities or by the formation of police forces, the spectacular element became increasingly symbolic, but still relevant. As Douglas Hay's study of eighteenth-century England shows, the parade of robed judges supported by military squads was an important demonstration of the symbolic violence of the law (Hay 1975; Spierenburg 1984).

The 'politics of law' is a summary term for the processes which have emerged and which continue to emerge, in myriad form, concerned with the contestation of techniques of law as governance. This is not about taking up positions or stances for or against 'the law' or some aspect of it. While we argue shortly that analysis of the politics of law involve assessments of the advantage and/or disadvantage of particular actors, the idea of taking some general position in regard to legal politics is not important to us; it is the technical relation between legal politics and governance which is of prime importance.

The politics of laws, as the contestation of laws as governance, is technically crucial as part of the perpetuating mechanism of law as governance. Whether law is involved in governing a business relationship, a passion killing, or a marriage, one technique of law as governance is always either being challenged by another or awaiting challenge. The form of the challenge – it may be that the dominant technique is being challenged for being too legalistic, too sloppy, too harsh, too lenient, etc. – is not as important as the fact that all techniques are always either being challenged or awaiting challenge, whether the challenging technique is fully formed, half-baked, or barely embryonic. The fact that techniques of law as governance are always either being challenged or awaiting challenge, the fact of the politics of law, is part of the perpetual cycle 'attempt at control – incompleteness (failure) – attempt at control' which has been identified.

In the governance of business relationships, the longstanding dominance of contract as a legal technique is always being challenged by alternative techniques, such as sovereign fiat (a single sovereign, like a king or an emperor, or sovereign body, like a military government, simply deciding at any time what business relationships should be like) or statutory regulation under company law statutes. The form of the challenge may be that the contract technique is too sloppy, or too restrictive, or too old-fashioned, or some combination of these or other judgements. The outcome of any challenge may be some sharing of dominance between contract and one of the challengers (currently statutory regulation), whether for a short or a long term. What is most important for our account is the fact of their constant challenge, that is, the fact that these challenges create contests over the governance by law of business relationships and thereby create legal politics.

The same can be said for the other two examples mentioned above, passion killings and marriages. In being governed by law, passion killings in the west are currently governed using techniques of modern policing and the application of modern criminal law. These techniques are always under challenge by alternative techniques, for example techniques concerned to achieve more immediate vengeance, techniques closer to blood feud or even Islamic law than modern western criminal law. Here the form of the challenge might be that modern criminal law is too slow and uncertain in seeking vengeance against passion killers. The outcome might be attempts to speed up the processes associated with modern criminal law, but whatever, legal politics is created.

In being governed by law, marriages in many western countries are currently governed using techniques concerned to ensure equity between the partners and with an eye to no-fault divorce, based on irretrievable breakdown, with just division of joint property and just division of care and maintenance of any children. These techniques are always under challenge by alternative techniques, for example techniques concerned to enforce patriarchal control of families, or to protect women from violent husbands and/or ensure 'the welfare' of any children. Here the form of the challenge might be that modern marriage and divorce law is 'too feminist', or 'too patriarchal', or unrealistic. The outcome might be modified marriage and divorce procedures, but whatever, legal politics is created, in this case often bitter, dramatic political contests as well as more mundane passive contests.

Resistance to law as governance is very much part of the fact that law can only ever help a social machinery run imperfectly or incompletely. Resistance to law is, like other forms of resistance, a counter-stroke to power.

We can sensibly speak of the challenge posed by sovereign fiat to the governance of business dealings by contract law as resistance to law as governance, as we can of the challenge posed by blood feud techniques to the governance of passion killings by modern western criminal law and of the challenge posed by patriarchal control to the governance of marriages by family law. We can even say that those individuals, groups and organisations involved in promoting these various challenges are engaged in resistance to the law. This should not be taken to mean that resistance drives the law in some conspiratorial sense. 'Resistance bringing down the law' or 'the law repressing resistance to avoid being brought down' are not central themes in our picture of legal politics. Politics and law operate in a much more technical, usually much more mundane, fashion than these somewhat romantic formulations suggest.

This is not to deny that the politics of law sometimes features exploitation and repression. At least two of our examples, passion killing and marriage, are rich with the possibility of fierce and bloody contests as well as passive and mundane contests. In talking of outcomes of advantage and disadvantage in discussing the politics of law as governance, we mean to cover the full range of outcomes from the mundanely passive to the spectacularly brutal. The contests over law as governance which make up legal politics can only, we are arguing, be studied at specific times. Analysis of legal politics involves a series of snapshots of ongoing contests. Attributions of advantage/disadvantage to the outcomes and actors involved should only ever be taken as attributions to do with a historically particular contest.

If analyses of the politics of law involved in an ongoing particular instance are undertaken over a long period of time and continue to reveal outcomes whereby one actor or group of actors remains in a situation of disadvantage, it is reasonable to understand this long-term disadvantage as structured oppression (even 'brutal oppression'). If such analyses continue to reveal outcomes whereby one actor or group of actors remains in a situation of advantage, it is reasonable to understand this long-term advantage as structured exploitation (even 'brutal exploitation'). Of course, our caveat from the previous chapter applies: while we can safely say that those instances when it is reasonable to assess technical advantage and disadvantage in more emotionally charged terms are those where the analyst considers that the outcomes deemed to be in disadvantage or advantage have no chance of being otherwise in the foreseeable future, such judgements are by no means easy to make. Political developments have a nasty habit of being much faster than those analysing them recognise.

To take the examples of passion killings and marriage, we cannot safely say that analyses of the legal politics involved reveal a clear, long-term advantage or disadvantage in either case. In the case of marriage, at least in the west, 20 years ago an analysis would have had to reveal structured oppression of women in the outcomes of the contests between the dominant techniques of the legal governance of marriage, especially the provisions for divorce, and alternative techniques concerned with greater equity for women. But in the last 20 years the situation has changed enough that any legal political analysis, while it may still assess patriarchal techniques of governance to be in a situation of advantage, would have to be uncertain as to whether the outcome of any contest is a 'long-term' outcome. In the legal politics of the governance of passion killings, no analysis in the last 50 years, at least in the western world, could have reasonably attributed long-term advantage or disadvantage to a particular outcome, such has been the volatility of contests between liberal techniques of rule of law and more traditional techniques concerned to achieve more immediate revenge.

It hardly needs saying that the law as governance constantly comes into contact with the darker side of resistance, what we called in the last chapter, in trying to capture Foucault on this point, the imperative to resist. In the last chapter we discussed Katz's account of the irrational element involved in many crimes. The imperative to resist, as we call it, is a key factor in the perpetual incompleteness/failure of the law's attempt to govern crime.

Sticking with our legal examples used in this chapter, the imperative to resist plays a role in the legal governance of business relationships, marriages and, of course, passion killings. We have already seen the way in which passion killings involve the sudden and dramatic urge to righteous slaughter on the part of the killer and discussed the impossible challenge this technique of self-governance throws up to legal techniques of governing killing. This example is fairly straightforward, even in regard to the fact that legal techniques of governance may prevent killings in 99 per cent of instances when passions rise; 1 per cent is sufficient to make our point about the role played by the imperative to resist in the incompleteness/failure of legal governance.

The legal governance of business relationships is far more mundane. Passions rarely rise to point involved in passion killings. Yet passion certainly plays a part in contract negotiations and enforcement, though much more in line with Bailey's *The Tactical Uses of Passion* than Katz's *Seductions of Crime*. The imperative to resist is involved in this example in that mysterious way of 'things never quite working' we referred to in the previous chapter, a meeting going awry in ways that no single participant anticipates or aims for, components not arriving on time despite the best procedures, contract negotiations ending up in court despite goodwill on the part of all parties involved, etc.

The legal governance of marriage is a mix of the mundane and the spectacular when it comes to the role of the imperative to resist. Breaches of legal governance by this imperative occur mundanely in events like inequitable division of marital property; irrational urges whereby one partner mistrusts the other and tries to maintain dignity via petty battles over who owns what are very common pieces of evidence that the legal governance of marriage to ensure equity is always incomplete/failing in the face of the imperative to resist. Sometimes the irrational urges involved in resisting the legal governance of marriage spill over into the realm of the spectacular. Violence by one party against the other, usually the man against the woman, is the result. This spectacular resistance to governance itself occasionally escalates as the imperative to resist that legal governance designed to ensure equitable marriages suddenly becomes the imperative to resist that legal governance designed to prevent passion killings.

Principle 3

Law as governance always involves knowledge.

As with governance in general, knowledge is used to select objects for legal governance and knowledge is used in the actual instances of legal governance. There are some important clarifications and qualifications necessary as we extend this aspect of our theory to law as governance. Here again we confront a paradox: that while knowledge is used to select objects for legal governance, the objects of legal governance are only ever known through governance. We stress the difference between this formulation and that we offered for governance in general in the previous chapter. Here, we are allowing that objects of legal governance may be known through non-legal governance. Again, we turn first to the Althusserian notion of always-alreadyness to overcome the effects of this paradox.

Objects of legal governance, like other objects of governance, are 'always-already' there, they have a past, possibly with beginnings, but they do not have origins in the sense of a determining genesis. When we speak of the legal governance of business relationships, marriages and passion killings, we do not propose that we, or any actors involved in this governance, must know the origins of business relationships, marriages and passion killings. While we or they may know or discover many facts about these objects, we and they (and here we can say 'the law' as part of this 'they') treat them as always-already there. Whether it is theorists like us, police officers, lawyers, judges, or law makers is not important; contracts can be managed, divorce proceedings undertaken, prosecutions made against passion killers and any number of examples formulated without recourse to knowledge of the origins of objects beyond their legal governance. Legal governance involves the cycle 'attempt at control – incompleteness (failure) – attempt at control'; it, like governance in general, is a cycle without end and a cycle without origins.

In overcoming the effects of this paradox, we assert that legal governance is always more important in social life than the known objects governed by law. Governing a known object by law is always more important for us than the knowledge of the object despite the usually assumed centrality of knowledge.

At a basic level knowledge works for legal governance in a seamless, invisible process. Knowledge selects objects for legal governance by invisibly posing and answering questions like, what is a human life? (allowing legal governance access to passion killings), what is human economic activity? (allowing legal governance access to business dealings) and how do humans traditionally partner one another? (allowing legal governance access to marriage). The objects involved are governed in that the basic acknowledgement of existence which knowledge performs involves a very basic attempt at control or management.

In line with this primary role for knowledge in legal governance, we could not know objects if they were not always-already governed, we could not be writing about them if they were not. Despite its central place in our theory of governance, this primary role for knowledge is not as important for law as governance as it is for governance in general. While the primary level provides objects for legal governance in the manner just described, this level is not as active in legal governance as it is in governance in general, because the secondary level comes into play far more quickly in instance of legal governance. It does this mainly in the distinctive form of 'legal knowledge'.

On the secondary level, where knowledge is used in the actual instances of governance, knowledge is used in choosing and implementing techniques of governance beyond the basic 'acknowledgement of existence' technique. This is where our new sociology of law as governance begins to ask, 'how?' (and where, as we hinted above, 'legal knowledge' becomes a sound general beginning to any answer).

On this secondary level, knowledge is used to impose more control or management. It involves both simple, definitional legal knowledge – taking a human life is a public wrong, human economic activity needs consistent regulation, traditional human partnerships, especially for the purpose of procreation, include marriage, etc.– and more complex legal knowledge – the evidence required to prosecute a passion killer might include a *mens rea*, contracts covering the supply of component parts should include clauses about delivery times, a divorce hearing should involve a notion of irretrievable marriage breakdown. This is the ground of law as governance. Complex and simple forms of legal knowledge are used to select some techniques of legal governance over others and to implement the chosen techniques in the attempts to impose legal control or management on the object or objects concerned.

Most legal knowledge involved in law as governance belongs in the range called rational knowledge, but this is not to deny that irrational knowledge is also involved. Indeed, we can make the same point we made in regard to governance in general: each technique of governance may include some irrational and some rational knowledge.

Consider, as a first example, the techniques of investigation used in prosecuting a passion killer. Some simple legal knowledge, like the knowledge that passion killing is a crime if successfully prosecuted, is involved alongside more complex legal knowledge to do with, say, properly gathering forensic evidence. Both the simple and the complex legal knowledge here is quite rational, but irrational knowledge to do with revenge may inform the techniques of investigation at any stage, weaving its way through the rational knowledge, sometimes supporting it, sometimes undermining it. As well, or perhaps as part of this irrational knowledge to do with revenge, knowledge about the role played by demons or devils in passion killing may informally inform the techniques of investigation. This knowledge is usually labelled irrational in modern western legal systems.

Even in more mundane settings, like the legal governance of a business relationship by contract, techniques of legal governance combine knowledge in a similar way. The contract technique of legal governance explicitly involves very rational knowledge to do with individual maximising behaviour, the behaviour of *homo economicus* of liberal theory. But this is not all it involves. Implicitly it also involves great faith in the possibility of order, which can well be held in the form of religious knowledge or even something approaching magic, both of which are on the irrational end of the spectrum of the sociology of knowledge. It may also involve (quite sensibly from our point of view) irrational fear of things going wrong, of some force or other intervening to throw spanners into the machinery of contract.

So, while legal knowledge appears super-rational, we are contending that in its central role in legal governance, it is nearly always a mix of rational and irrational knowledge.

It is necessary that we rehearse our three connected points about Foucault's special influence on our thinking about the role of knowledge in governance specifically in regard to law as governance. First, the knowledge which is involved in legal governance is always available knowledge. Second, the knowledge which is predominantly involved in modern legal governance is very much part of the formal knowledge groupings known as the human sciences. Third, statistical knowledge is increasingly important to legal governance.

The knowledge which informs the legal governance of marriage, for example, is made available by the operations of institutions to do with human partnerships and human procreation. In the late twentieth-century western world these institutions include education, counselling, therapy, family, friendships, peer pressure, church, court (especially family court), medicine, midwifery, popular health books and magazines and other popular books and magazines (especially to do with romance, marriage, child-bearing and child-rearing). These institutions make this knowledge available to the law for its governance of marriage at the expense of other knowledge. For instance, it is unlikely that witchcraft will be available to the law in the late twentieth-century western world as it goes about governing marriage, though it would be difficult to be so certain about the exclusion of astrology. Modern western legal governance of marriage, in line with the role of the above-mentioned institutions, is heavily informed by the human sciences. The law is now dominated in its approach to marriage, by knowledge given to it by psychology, sociology, sexology (especially as they relate to the management of 'healthy populations', however that may be defined), as well as biology and medical science. Of course, these various human sciences are heavily indebted to statistics as they go about the production and promotion of knowledge about human relationships and human procreation which laws governing marriage cannot resist (the statistics are central in calculations to do with how a 'healthy population' is achieved).

Principle 4

> Law as governance is always social and always works to bind societies together (which sometimes, ironically, involves social division).

For law as governance, our two senses of 'social/society' are also important. In the traditional sense of 'society/social', law is a part of that which pre-exists individuals in consideration of the collective actions of individuals. Again, Durkheim is the key figure, supplemented by Althusser's notion of the always-already. The law, as part of society, always-already pre-exists the individuals, in their collective and individual actions, who are the subjects of law. Law is always-already there as part of society. There is no point searching for the origins of law in society. One can find a past, one can find beginnings, but one cannot find the origins of law in society, in the sense of a genesis. Our sociology of law as governance is the study of how – the exact detail of how – law in society is always-already there.

For our second sense of 'society/social', the Foucaultian sense, law is a part of governmentality, part of the form of government unique to the modern world. Law in society in this sense is an invention. It was invented as a definite category of the government of nation-states and regions in the nineteenth century and further developed as such a category throughout the twentieth century. In this Foucaultian sense, law in society is a modern western phenomenon (though one which has been exported to non-western nations). This Foucaultian understanding of law in society is, in important respects, part of the inspiration for this book, certainly for our formulation of a new sociology of law as governance.

To continue with our account of this aspect of 'law in society', it is a relatively new conjunction of concerns about populations, their longevity, their education, their ethical behaviour, their children, in short, their 'health', in the broadest sense of the term. The stimulus for the invention and development of this sense of law in society was and continues to be some dramatic advances in insurance technology. In line with some developments in probability theory, it became possible to think about, to make definite calculations about, to make policy about, to make provision for, the 'health' of larger and larger numbers of people and the law became a central part of this process, a sort of guarantee of its possibility and a part of its execution. While governments were able to think about these concerns before, even to make policy about them, and to use the law in so doing, the distinctive new ingredients were the fine detail of the calculations and a device whereby governments could attempt to provide financially for the future of the population on the basis of these calculations. Law became a detailed means of both expressing and attempting to implement more and more detailed policy in line with the new intensive mode of government. In this way, crucially, law became an arm of calculating government.

This Foucaultian sense of law in society dovetails with the traditional sense. The fact that law in society is always-already there is boosted by the emergence of a new field of government, with a concomitant new field of law, around the social. From the late nineteenth century 'society' attracted an enormous amount of new attention which was accompanied by new attention to 'law in society' which in turn encouraged much more thinking, talking and writing about its always-alreadyness. This situation continues as we approach the end of the twentieth century. The new fields of law which were made possible by this new field of government, child protection law, family law, social security law, workers' compensation law, etc. and, by extension, as 'society' became a wider field of calculation and concern and as more intensive governmental techniques became available, such fields as environmental law and new, more intense forms of company law, taxation law and international law, all now seem 'obvious' and 'natural' as legal fields. The boom for government has been and continues to be a boom for law.

All instances of law as governance are social in the traditional sense, though not in the Foucaultian sense. It will be remembered that the objects of law as governance are always-already known through governance. In this way, law as governance is always social. In any instances of law as governance, the objects and the techniques are made available by society, they are always-already available. Whether it is a child custody dispute being governed by law or an oil spill, this is the case. Both the child custody dispute and the oil spill are always-already available to the actors involved, they are socially available objects, they have no existence beyond society. Whatever techniques of law are applied in these two instances – conciliation, restraining orders, negotiation, prosecution – they are always-already available, they are socially available techniques, they have no existence beyond society.

For the Foucaultian sense of law in society, only some objects of law as governance were initially made available by society – those invented or modified as part of the invention of the social (such as the dysfunctional child, the threatened environment and the unsafe workplace) – and only some techniques of law as governance were initially made available by society – those invented or modified as part of the invention of the social (such as compulsory education, safety inspections and environmental standards). Fairly quickly, however, the objects and techniques which were only available because of the invention of the social as a new field of government through law merged with and modified older objects and techniques such that now one cannot sensibly separate the traditional sense of 'law in society' governance from the Foucaultian sense. Nonetheless, as we turn to the fact that law as governance is always working to bind societies together, we can acknowledge that as with governance in general, this principle is more Durkheimian than Foucaultian.

While law in society is always-already there, this does not mean, of itself, that it is always-already there strongly enough to bind actors into it. Being always-already there might entail, for example, no more than the pre-existence of the idea that taking a human life is problematic, for the legal governance of passion killings, or the pre-existence of the idea of patriarchal arrangements for organising human procreation, for the legal governance of marriage. It does not necessarily entail actors being bound to one another around these objects. For actors, whether individuals or organisations, to be bound together, law as governance has to work to strengthen the always-alreadyness of law in society, to add this extra dimension of boundedness to the traditional meaning of law in society.

Law as governance is perpetual in being always incomplete or failed. So, law's part in binding societies together is always incomplete or failed. In the Durkheimian account of law as a binding mechanism, we can see law working closely with the mechanisms identified and discussed in the previous chapter, morality, community, communications and physical structures and sacred rituals.

The law supplements morality in its task of unifying individuals and organisations around particular themes of right and wrong, like the theme of the value of human life. It should be remembered that the content of the morality does not matter for this account. The law may support Christian morality, pagan morality, or whatever, to the extent that it can quite usefully back up the flexible moral code needed to encourage and/or allow the taking of human life at some times (times of war, executions, etc.) while condemning it at others. What is important is that the law helps guide/coerce/encourage/induce individuals towards certain ways of doing right and away from certain ways of doing wrong, even where flexibility is required, as in the above example, and this helps to bind them into a society. What is also important is that the binding never works completely; at least some individuals and organisations slip through the nets of morality which the law helps to set, for shorter or longer periods.

The law supplements those techniques of intimacy, continuity and cohesion which make up 'community', as they operate alongside morality in attempting to unify individuals and organisations around themes of right and wrong. For example, the law helps produce communities of locality or ethnicity by restricting immigration, criminalising vagrancy, etc. Here again, the content of law does not matter; it is the operation which matters. Here again, too, the law never works completely in supporting community as a binding mechanism; no laws, not even those associated with apartheid or Nazism, have ever produced a completely isolated community of locality or ethnicity.

The law helps communications and physical structures as they jointly perform their binding work in this Durkheimian picture of social binding. The law helps them as they contribute to the morality-community mix in both 'simple' and 'complex' societies. In 'simple' societies, with relatively small numbers of individuals and organisations (or possibly no organisations) spread over a relatively small area, the law helps regulate and control immediate verbal communication (via taboos on certain words, for example) and any limited forms of stored communication and communication over distance (via legal recognition of any designated formal communicators, especially in their capacity as communication storage devices, like storytellers or dancers). In such societies, the law helps the physical structures side of the combination it forms with communication by, for example, legally marking certain structures, particularly religious structures, as special. While the law-communications-physical structures nexus is never complete or completely successful in its binding work in these societies, it has less binding work to do than it does in societies with relatively greater numbers of individuals and organisations and/or with relatively greater distances between them.

In these more complex societies, with their much more complex communication devices and physical structures, the law aspect of the law-communications-physical structures nexus still has to enforce taboos on certain communications content, to provide formal recognition to designated communicators (telephone companies, universities, television networks, etc.) and to mark certain structures as special (schools, hospitals, prisons, churches, government offices, business centres, shopping centres, etc.). But it does much more as well. For example, it regulates and polices ownership systems for communicated ideas and words, it backs up the taboos on certain communication content with libel laws and censorship provisions, it regulates competition in communications industries, it regulates to ensure the spread of communications technology, it regulates complex transportation forms and patterns and it regulates the complexities of modern urban spaces. Little wonder, then, that incompleteness/failure of law's binding work is more apparent in regard to these mechanisms in complex societies.

The law's role in the over-arching binding mechanism of the sacred has already been touched upon. The law is crucial to the project of marking some objects and practices as sacred and others as profane. Law helps in binding individuals and organisations around the sacred and against the profane. Church law is the most obvious example, whether Christian, Islamic, Jewish, or whatever. However, as the sacred operates across all spheres of social life, not just religion, it is important to note the work the law does in non-religious settings to regulate the distinction between the sacred and the profane. For example, law regulates this distinction with regard to the objects and practices of the celebration of military achievements and sporting achievements, of the provision of medical and educational services and of the provision of formal government. It is difficult to imagine a modern western society without law policing the sacredness of war memorials and rituals, sporting venues and events, hospitals, doctors and 'proper' health care, schools, teachers, universities and education, government buildings, parliaments, politicians and public bureaucratic procedures.

Law as governance helps bind societies together, never totally or completely successfully, even in cases which appear at first glance to be dysfunctional for binding. Durkheim is famous for this contention, especially in regard to crime. Of course there can be genuine anti-law activities where there cannot be genuine anti-governance activities short of death. Law has its limits, as we have been at pains to argue, especially through our Weberian definition of law; the rule of law has to be established and constantly maintained. In this light, it is easy to see that our account of law as governance, especially in regard to our argument that law as governance is perpetuated by its incompleteness/failure, might more accurately be called dysfunctionalism than functionalism.

Conclusion

The content of our new sociology of law as governance has been outlined in terms of the four principles of governance. Our Weberian definition of law is central to these concerns, in limiting the field for a sociology of law as governance. Admittedly it is a large field, but it is a limited field nonetheless. Now that we have a field, we need some rules for the game. The rules, in the spirit of Durkheim, are rules of sociological method. It is to these that we now turn, but with Foucault in our eyes as much as Durkheim.

6
Method Principles for the Sociology of Law as Governance

Introduction

Foucault and Durkheim come together in this chapter, but not in the sense of trying to force Foucault's genealogy into the framework of Durkheim's *Rules of Sociological Method*. That would not, in our estimation, be particularly productive, even if possible. Indeed, we attempt no explication of either theorist's work here. We have done all the explicating of Foucault's work we are going to do in this book and explication of Durkheim's work is outside its scope. Rather, we put Foucault and Durkheim together by giving our Foucaultian project some Durkheimian methodological spirit; we draw on insights by both writers in doing this but mainly on Durkheim's insights.

We have discussed the fact that Foucault refuses to be pinned down on matters of method. We find this fascinating for its philosophical ramifications but decidedly unhelpful for our task of outlining a new approach to the sociology of law, even though our new approach is definitely Foucaultian. We offer five explicit principles of method with the same conviction displayed by Durkheim (and Weber); we are certain that a strong sociology needs a set of explicit methodological principles.

The four method principles of the sociology of law as governance

Principle 1

> The sociology of law as governance works to compile social facts in a genealogical manner.

The sociology of law as governance goes about its business compiling what Durkheim calls social facts, but compiling them in the manner of Foucault's genealogy. We deal with the Durkheimian component of this formulation first (we draw on Durkheim 1964; Carrithers et al 1985; Collins 1985; Nisbet 1965).

As we saw in the previous chapter, all law as governance is social governance involving social actors, individuals and organisations bound into a society. The sociology of law as governance follows Durkheim in detailing the actions of these actors as social actions, as 'things' which happen through actors interacting socially, that is, always-already interacting. For Durkheim, sociology is the science of these social actions (the French term *fait social*, usually translated as 'social facts', more literally means 'social doings'). The actions of actors are never detailed in other than social terms. Phenomena often attributed to non-social (usually individual) realms, phenomena like motives and beliefs, are always detailed in terms of their social make-up, in terms of their dependence on the always-already interaction of actors. In this way, supposedly personal phenomena are detailed in terms of social processes of person formation (a point made most forcefully by Durkheim's immediate followers, especially Marcel Mauss (1985)).

Consider the example of abortion law.[1] Governing abortion by law is social governance. All actors involved, women, doctors, police, politicians, hospitals and courts, are social actors, that is, are individuals and organisations bound into a society. The sociology of law as governance, in addressing abortion law, details the actions of these actors – getting pregnant, deciding to terminate a pregnancy, medically helping the termination, or medically advising against the termination, investigating an abortion, or making an arrest, or deciding against making an arrest, helping to frame laws against abortion, or opposing such laws, or helping to frame laws to allow women access to abortion, providing the services for a medical termination, prosecuting those involved in an abortion, or deciding not to prosecute – as social actions. This is to say, the sociology of law as governance details these actions as 'things' which happen through actors always-already interacting; the possibility of getting pregnant, deciding to terminate, helping medically to terminate, etc., always-already pre-exist the actual getting pregnant, deciding to terminate, because actors always-already interact. The sociology of law as governance is the study of such social actions. The motives and beliefs of the women, doctors, lawyers, judges, politicians, police, hospital employees, involved in abortion, like all other supposedly 'personal' phenomena, are detailed in terms of their social make-up, in terms of their dependence on the always-already interaction of actors. The sociology of law as governance examines the ways in which this always-already interaction makes possible the beliefs and attitudes that sex is desirable, that pregnancy is undesirable, that doctors should help terminate pregnancies, that doctors should not help terminate pregnancies, that police should prosecute, that police should keep clear, that anti-abortion laws are necessary, that they are unnecessary, that special laws are necessary to facilitate abortion, etc. This brings us neatly to the Foucaultian component of this methodological principle.

We discussed Foucault's notion of genealogy on a couple of previous occasions, drawing attention to its stress on identifying conditions of possibility and to its capacity to disturb the obviousness of the present. Adopting Foucault's genealogy means the sociology of law as governance involves the compilation of social facts in order not to make sense of the present, but as a constant demonstration that the present is nothing special, that it is what it is, a collection of contingencies, in some ways unique, in some ways the same as other eras. In line with this, the sociology of law as governance involves the compilation of social facts from different societies and different eras as well as from the 'home' society of the sociologist and from the present. We stress, however, that it is not legal history, in the sense of using facts about the past to make sense of the present, and it is not legal anthropology, in the sense of using social facts about other societies to make sense of the 'home' society. While this point is particularly Foucaultian, it is not foreign to Durkheim.

With this Foucaultian component added, this methodological principle means the sociology of law as governance approaches abortion law with the aim of compiling social facts about abortion to establish their status as social facts, that is in answer to the central question, how are these social facts possible, what are their conditions of possibility? The sociology of law as governance investigates the actors involved in abortion law (women, doctors, police, politicians, hospitals) as contingent social actors. It details the involvement of these actors to highlight their contingent participation, the conditions of possibility of their participation; women are involved in line with the strong operation of beliefs that only women are the 'natural' bearers of children (even in an era where other technological possibilities exist), doctors are involved in line with the victory of western medical expertise over other forms of health expertise, police are involved in line with the capacity they acquired only late last century and early this century as key actors in the governance of public morals. The sociology of law as governance investigates the actions of these actors – getting pregnant, deciding to terminate the pregnancy, helping to frame laws against abortion, or to make abortions more accessible – as contingent social actions. It details these actions to highlight their contingency, their conditions of possibility: getting pregnant is made possible not just by having sex but by the operation of certain attitudes and practices towards sex, to contraception, etc.; deciding to terminate a pregnancy is made possible by the operation of certain attitudes towards it and certain technologies which make it available, etc.; helping to frame laws against abortion is made possible by the operation of a definite legal system and understandings about the appropriate involvement of that legal system in matters of the morality of human reproduction, etc.; helping to frame laws to make abortion more accessible is made possible by the same factors but with the crucial addition of a concern with individual rights, etc.

This arm of our methodological picture is clear enough. For those readers concerned that the intellectual processes involved here have no solid base, that social facts are compiled to highlight contingency, conditions of possibility, at the same time that these conditions are themselves contingent, that they in turn have conditions of possibility, we can only acknowledge their perspicacity. This arm does indeed have no solid base. We provide a sort of base through the other principles, but it is not solid in the traditional sense; it does not pretend to offer any guarantee as to its truth. Our Foucaultian/ Durkheimian universe has turtles all the way down.[2]

Principle 2

> The only tools employed by the sociology of law as governance are attention to detail and careful generalisation.

At least the first half of this principle has been already introduced. By 'attention to detail' we mean the presentation of details of genealogical social facts with great care and exactness. The sociology of law as governance is concerned to compile more and more details of instances of law as governance, with as much care and exactness as possible, in line with the four content principles outlined in the last chapter. This is to say that the sociology of law as governance involves the detailing of more and more instances of the 'attempt – incompleteness/failure – attempt' cycle of the legal management of things, always featuring different techniques (including possibly different technologies of power) and political contests around these techniques, always involving the use of knowledge and always being part of society, part of attempts at social binding (in the traditional and Foucaultian sense of society). Of course the attention to detail must also be in line with the four method principles of this chapter.

Turning now to the second half of this method principle, all eight principles (four of content, four of method) of the sociology of law as governance are established, careful generalisations. They are generalisations reached over a long time by induction from a great deal of careful detail. In using this set of generalisations as the basis for the subdiscipline's work, it has to be said that at least some deduction is involved in the sociology of law as governance. While we take induction to be a more likely indication of careful research, neither can be said to guarantee it. Indeed, nothing can be said to guarantee it. The most a discipline or subdiscipline can do, if it has attention to detail at its heart, by way of ensuring the care which defines this research tool, is to stress the importance of a careful approach in all its institutional dealings with its practitioners, through its books, journals, conferences and university departments.

In studying the legal profession, for example, the sociology of law as governance compiles more and more details of the ways the profession is governed and the ways it contributes to legal governance with as much care and exactness as possible, in line with the four content principles outlined in the previous chapter. This is to say that the sociology of law as governance details instances of the 'attempt – incompleteness/failure – attempt' cycle in regard to the legal profession's governance of court procedures, itself, government actions, conveyancing, etc. and instances of this cycle in regard to the governance of the legal profession by governments and some others (perhaps private industry), with attention paid to the techniques of governance involved (like professional codes of ethics, government legislation, internal manoeuvring, reliance on traditions, etc.) and to the political contests around these techniques (like contests between governments and the profession about government regulation in the face of the profession's insistence that internal codes of ethics are sufficient, contests over the profession's attempted monopoly on conveyancing and on court procedures, contests over a particular business deal, etc.). The detailing always covers the use of knowledge (the ways the profession and governments deploy knowledge in their contests over regulation, the ways knowledge of tradition is used by the profession to protect itself, etc.) and always covers the social binding dimensions of the legal governance in question (the ways the profession constitutes itself as a community, thereby excluding others, the role of the profession in binding actors around the idea of the rule of law and due legal process, etc.).

It is on this basis that we discuss the production of new generalisations. As more and more details are compiled, new generalisations arise. They may concern any aspect of the content principles elaborated in the previous chapter or method principles elaborated in this chapter, they may even lead to new content principles or new method principles. It is impossible to predict what new generalisations will arise and it is improper to attempt to do so; the sociology of law as governance is restricted to attention to detail and careful generalisation; prediction is not a proper part of the subdiscipline's work. What can be properly said, as a careful methodological generalisation, is that all intellectual disciplines produce new generalisations; it is historically part of their work.

The examples we use in outlining the sociology of law as governance are generalisations arrived at in the manner discussed above. When we talk of the legal governance of passion killings, marriages, business dealings, abortions and the legal profession, we are of course making use of generalisations, as we indicated, in line with our content principles and method principles. It may well be argued that our generalisations are not as careful as we are arguing generalisations should be. Admittedly we do not provide the detail necessary completely to rebut such a charge; this is not our project in this book (we are outlining a new subdiscipline here and just as the plans for a new car plant can be sensibly outlined without the planners having to lay out every component of every car ever built, or having to prove they personally can build cars by hand, so an outline for a new subdiscipline can be sensibly offered without the planners having to lay out every component of every generalisation ever built, or having to prove they personally can build generalisations by hand). Nonetheless, we feel we have a strong argument against this charge simply by dint of the fact that no generalisation we use contains even a hint of the reckless generalisations which have plagued sociology and other social sciences for most of this century: reckless generalisations to do with, for example, class, race and gender.

This discussion indicates our final point for this section: there is a fine line between careful and careless generalisation. As we said in a slightly different context, the appropriate amount of care cannot be guaranteed. The most that can be done is to stress the importance of a careful approach in all institutional dealings. Here, we suggest that a device for encouraging care, to be used in these dealings, is a rule that the benefit of the doubt in regard to generalisations must always favour further detail. In other words, if there is any doubt that a generalisation has not been reached with due attention to detail, it should not be accepted as a generalisation, further detail should be compiled before another generalisation is attempted.

Principle 3

> The basic production work of the sociology of law as governance must always be distinguished from the uses of its results.

The attention to detail and careful generalisation which make up the subdiscipline are not at all the same thing as the uses to which the details and generalisations are put. We know, especially from Foucault's work, that the products of the social sciences are used for many and varied purposes, especially governmental purposes, not all of them noble.[3] The sociology of law as governance can take some steps to limit the use of its details and generalisations, but no more than any other social scientific endeavour; these steps, it must be recognised, are minimal.

For example, the sociology of law as governance may compile many details about the legal governance of abortion in line with its principles.

It may even come up with new generalisations which can act as new principles. These products are the result of the basic work of the sociology of law as governance. Yet they might be used by any number of actors not connected to the sociology of law as governance such as women contemplating abortion, politicians, political activists, police, judges, lawyers, doctors, hospital administrators. Furthermore, they might be used for a variety of purposes – to help in deciding whether to have an abortion, to help frame legislation banning abortions, to help frame legislation making abortions more accessible, to help an anti-abortion political campaign, to help a pro-abortion political campaign. The sociology of law as governance can take some steps to limit some or all of these uses of its products. It can, through its institutional dealings (university departments, books, journals, conferences), encourage governments to pass laws limiting the use of its products to, say, doctors and hospital administrators (the irony of encouraging legal governance to protect the study of legal governance is overwhelming). It can encourage other institutions to use internal procedures (like lawyers' and doctors' codes of ethics) to limit the use of its products; and it can use existing legislation to do with intellectual property and defamation to limit the use of its products. It can do little more than this and these steps are minimal. In modern western societies, with their elaborate knowledge production, storage and reproduction technologies, the spread of knowledge products like those of the sociology of law as governance is almost impossible to contain.

The sociology of law as governance cannot survive as an independent subdiscipline if it does not maintain this strict distinction between, on the one hand, attention to detail and careful generalisation and, on the other, their uses. The definition of governance at the heart of the subdiscipline is so wide it is always potentially under pressure from a huge variety of governmental concerns interested in directing its production activities. We glimpsed this in the above discussion of the legal governance of abortion. The situation is similar no matter what instance of law as governance is being studied whether it is crime, police, legal profession, judiciary, contract, company law, or marriage. The distinction featured in this method principle provides at least some protection against the encroachments of the users of the subdiscipline's products on its production mechanisms, though of course, as we have stressed, it does not provide a guarantee of independence.

Within this principle we can see an important difference between the methodology of the sociology of law as governance and Durkheim's methodology (and indeed most other established methodologies of science and social science). This difference centres on the notion of the uses of the knowledge products. For the sociology of law as governance, all uses are uses, no matter what name they are given and no matter which agency or actor is doing the using, including other knowledge-producing agencies and even the sociology of law as governance itself. Crucially, uses are uses and must be separated from the basic knowledge products even where the uses are called 'causes' or 'explanations' or 'predictions', whether by some other knowledge-producing agency, by some other actor, or by the sociology of law as governance itself. By our account, the sociology of law as governance is properly restricted to the production of details in line with careful generalisations and occasionally new generalisations (producing new careful generalisations is the only process in which the subdiscipline can properly use its own products). It has no business calling some of its products 'causes', 'explanations', or 'predictions'; these are definitely uses.[4]

So, in this way, the sociology of law as governance does not produce 'explanations' of crimes, police actions, the legal profession's conduct, judges' behaviour, business dealings, contract undertakings, marriage practices, abortion, etc. It does not seek to, or even accidentally, uncover 'causes' of crime, police action, the legal profession's conduct, etc. and nor does it seek to provide predictions about future crimes, police actions, etc. It must be remembered that the sociology of law as governance addresses 'how' questions, not 'why' questions.

This is not to say, we stress, that causes, explanations and predictions will not be offered on the basis of the subdiscipline's work. How could they not be? So much intellectual effort in the twentieth century has been directed, almost obsessively, to causes, explanations and predictions. We contend that causes, explanations and predictions are uses of the subdiscipline's work, not the work itself. Causal, explanatory and predictive thinking are discouraged in much the same way as Foucault's work discourages them.

Principle 4

> The sociology of law as governance is a continuously reflexive subdiscipline.

The ground on which the subdiscipline's reflexivity exercises itself has already been laid: the sociology of law as governance is reflexive in that among the instances of governance it studies are the uses to which its own products are put, including the uses to which it itself puts them.

For example, in studying instances of the legal governance of company takeovers, the sociology of law as governance focuses on the details of the 'attempt – incompleteness/failure – attempt' cycle (the techniques of boardroom lobbying, government regulation and the politics involved, such as contests to impose government regulation on boardroom lobbying, etc.). These details may or may not be linked to other details to produce a new, careful generalisation. Throughout this process, indeed as part of it, the sociology of law as governance must compile details of the uses to which its details are put, perhaps by the actors involved (company directors, government regulators, journalists, etc. may all find the subdiscipline's details useful), perhaps by other branches of social science (economics, political science, even history, may find these details useful). Of course this must also involve, wherever and whenever it happens, compiling details of the uses to which it, itself, puts these details. This may mean compiling details of the subdiscipline producing new, careful generalisations from its stockpile of details; this is the only proper use of its products we allow for the subdiscipline. Or it may mean compiling details of internal uses of the details which we regard as methodologically improper, but which we acknowledge as an inevitable part of any discipline's existence.

This is a crucial acknowledgement. It is perfectly consistent with our theory of governance. The methodological principles, including the principle restricting the subdiscipline's uses of its own products, are techniques of governance, techniques for governing the subdiscipline. This governance, of course, is always incomplete (sometimes to the point of failure). So we do not expect the methodological principles to work completely as governing devices. We are building into this principle of reflexivity an imperative for the sociology of law as governance to compile details of its own methodologically improper uses of its own products; of course we hold out no hope that this imperative works completely; it may even fail spectacularly. This point is extremely Foucaultian; Foucault is very fond of forcing disciplines to examine their internal use of their own products, especially where this means making them face up to the skeletons in their own cupboards.

The sorts of things we have in mind here, which are not always spectacularly dirty, we stress, are that some practitioners of the subdiscipline may try to establish a new generalisation as a new content principle while others resist on the grounds that not enough details have been completed; some practitioners may attempt to import generalisations into the subdiscipline from elsewhere as new content principles, perhaps to do with class, race or gender, while others resist, as well as the much rarer and more sordid instances whereby some practitioners try to force other practitioners out of the subdiscipline's institutions by denying access to journals or refusing tenure.

In this vein, the sociology of law as governance must be continuously aware of its own institutional nature. The practitioners must be aware that it is its institutions which set its limits, not some pure quest for knowledge and/or truth. It is precisely the institutional character of the subdiscipline (that is, its institutional governance and the politics of this governance) which determines which objects are addressed and the manner in which they are addressed. For example, it is the institutional arrangements (a university department and its relations with outside bodies, books, journal articles) and the politics associated with their governance (internal contests about how to run the department, contests with the university administration and with other departments in the university about how to run the department) which determine whether the subdiscipline addresses a particular object – whether the police, the legal profession, business dealings, abortion, crime, or whatever – and, crucially, how it addresses it (what details are relevant, which content principles are relevant, which method principles are used, etc.). Despite the form of this part of the book – detailed abstract principles – we are aware, and are suggesting that all practitioners within the subdiscipline must be aware if they are to be good practitioners, that the institutional governance of the subdiscipline and the politics associated with it determine the way the subdiscipline goes about its business, not a set of abstract principles working in the vacuum of scientific quest for true knowledge. We could not be judged good Foucaultians without an expression of this awareness.

In line with this, we stress that the subdiscipline can have no predetermined focus on the 'macro' or the 'micro', as both the objects and methods consistent with the use of these labels are subject to the politics of the field, to the governance of the sociology of law as governance. We have studiously avoided, in our principles, suggesting that any object is a 'natural' or 'obvious' object of study. We have been careful to point out that no point of method can ever be applied in a 'pure', institutionally uncontaminated, apolitical way.

Conclusion

The four method principles outlined in this chapter make the sociology of law as governance look remarkably similar to many nineteenth-century sociological ventures; not just those of Durkheim and Weber, but also those British and continental sociological projects which sought to map a social terrain and which saw little difference between statistics and sociology (Abrams 1968). Obviously the presence of a Foucaultian component in our methodological discussion, the importance of genealogy, renders this comparison somewhat inaccurate. Nevertheless, in terms of the spirit of methodology, the comparison is accurate. We are perfectly happy to be seen to be proposing the outline of a map. The social terrain we want the map to cover is that on which law is a crucial part of governance. The sociology of law as governance is very much a kit for mapping.

7
Conclusion: The Sociology of Law as Governance at Work

Introduction

This book has moved its focus from introduction to Foucault's work on law, through explication and criticism of this work, on to an outline for a new Foucaultian sociology of law. The first two parts of the book stand concluded and we use this conclusion to wrap up the third part. We offer a more detailed example of the work of our new sociology of law as governance than any presented thus far. We focus on a standard sociology of law topic, namely, the operation of modern western police forces. We discuss this object in terms of the eight principles of operation of our new sociology of law, but not separately. We have presented the principles separately for ease of exposition.

The police as a topic for the sociology of law as governance

In the day-to-day practice of the subdiscipline the principles do not, of course, function so neatly. The situation is much more confused as principles work their way in and out of one another and, as we have been at pains to stress, sometimes disappear altogether from the subdiscipline's agenda. As we are here presenting a conclusion to our outline rather than a complete study within the sociology of law as governance, we do not follow the principles on this complex journey (something which can only be done on a case-by-case basis anyway). However, to give more of a flavour of the way they work in practice, we discuss the subdiscipline's treatment of the police in terms of three working clusters of principles.

The first cluster contains the method principle which stresses the compilation of social facts in a genealogical manner and the content principle which stresses the way law as governance is always social and therefore always part of social binding mechanisms. The sociology of law as governance compiles details about the police in terms of their role in binding communities (which ironically involves dividing them) around particular codes of right and wrong and around localities or ethnicities, but always with an eye to disturbing the obviousness of the present of modern western policing. The subdiscipline looks to the past to help in this task. It finds (and here we are drawing on especially Foucault 1988b, but also Pasquino 1991; Chapman 1971) a somewhat different conception of the 'police', operating from the seventeenth through to the nineteenth century, from the one we regard as obvious in the twentieth-century western world (a 'plague of blue locusts'). We touched on this conception of police earlier in discussing governmentality.

The older doctrine of police operated alongside the doctrine of reason of state. While the doctrine of reason of state defined the principles and methods of government by states and marked it off from, for instance, God's governing of the world and the father's governing of the family, the doctrine of police defined the nature of the state's rational activity, its aims (especially the utilitarian aim of the happiness of the population) and 'the general form of the instruments involved' (Foucault 1988b: 73–4). The authors concerned with formulating this older doctrine of police (mainly German and Italian authors) understood police not as 'an institution or mechanism functioning within the state, but a governmental technology peculiar to the state; domains, techniques, targets where the state intervenes' (Foucault 1988b: 77). To sum up this older conception of the police: 'In short, life is the object of the police: the indispensable, the useful, and the superfluous. That people survive, live, and even do better than just that, is what the police has to ensure' (Foucault 1988b: 81).

The nineteenth century saw the big shift toward the more limited understanding of police we are familiar with today. This 'new police', as it is sometimes called, quickly came to focus on criminal activity and social order. Within these limits and as part of them, new techniques of surveillance allowed policing to become much more the work of specific 'forces'. It also allowed special or secret police forces to develop to deal with threats to particular governmental regimes; the policing of the 1848 uprisings across Europe and the 1871 Paris Commune stand out as early examples of this type of policing (Foucault 1988b, 1988c; Styles 1987; Chapman 1971).

Using the genealogical approach in this manner, to unsettle the obviousness of the police's role in twentieth-century societies, the sociology of law as governance can take a very broad approach to the social binding work of the police. It compiles details about all aspects of police operations (crime prevention, crime detection, traffic work, social order maintenance) in terms of the police's technical role in binding a local community around a code about which behaviours are right and which are wrong, a code about the destiny of one ethnic group as opposed to another. In this way, the sociology of law as governance is in a position to demonstrate connections between the old and new conceptions of police, that is, it compiles details of police operations in terms of their technical role in striving for a technically happy population, that is, a population that survives and strives for moral and physical health, however they are defined.

The second cluster of principles combines, on the one hand, attention to detail and careful generalisation with, on the other hand, the three generalisations which constitute the three other content principles. This is to say, the sociology of law as governance compiles details of the police in line with the following three careful generalisations: all instances of policing as legal governance contain elements of attempt and elements of incompleteness (which at times may be seen as failure); policing as legal governance involves power and as such involves politics and resistance; policing as legal governance always involves knowledge. We deal with each one in turn.

The notion of the 'attempt – incompleteness/failure – attempt' cycle in regard to policing has been raised several times, especially in discussing the legal governance of theft and of abortion. We mean by this that, as with all other instances of governance, policing (whether formal policing by a public police force or a private force, or even informal, occasional policing by, say, an insurance company) involves attempts to control or direct objects which have come to be the targets of policing. We know from the above discussion that these objects vary over time and place, even so we can sensibly talk about the policing of crime, morals and happiness as the policing of health, in its broadest sense. Whatever the particular object being legally governed by policing, the attempt at control or direction is always incomplete, thus perpetuating the governing process in the manner we have discussed.

What remains to be added in this case is the remarkably open and direct way in which failure leads to perpetuation, particularly in regard to the policing of crime. Using police forces to govern crime comes nowhere near complete governance. It is fair to say that since crime statistics began to be systematically kept towards the end of the nineteenth century, by this important and widely used measure, police forces around the world have failed abysmally to control or direct crime in the way they are explicitly charged with doing. This has produced little or no surprise on the part of most actors involved in the establishment and maintenance of these forces and produced no sustained policy suggestion that police forces be scrapped as a means of governing crime. Quite the reverse: their failure has consistently led to more and more resources being directed their way. It would be hard to think of a better example: the failure of policing as legal governance leads to more policing as legal governance.

Policing as legal governance involves particular techniques of power to do with increased surveillance and restrictions on the use of violence. We glimpsed the surveillance aspect earlier in some of Foucault's remarks about the shift which occurred in policing in the nineteenth century. We understand this term in a technical way, not in terms of some romantic threat to individual freedom. Policing involves gathering information and using it in various ways, whether to promote traffic flow, apprehend speeding drivers, solve crimes, secure more funding, achieve promotion, find missing persons, control a strike, protect a political regime, or whatever. We use the term 'surveillance' simply as a shorthand for the plethora of techniques of gathering and using information available to the police.

This is not to deny the importance of the second half of our 'police power' formulation concerning the use of violence, but rather to put it into perspective. Certainly the fact that the police use violence as a part of being a police 'force' is important. However, without the fairly mundane information techniques we summarised as 'surveillance' the potential for violence would be too random for policing as legal governance to have any potential as the sort of governance modern government requires. Indeed, police violence is always subjugated to police bureaucratic information techniques as a key feature of modern policing (even in cases where the procedures fail to govern police violence in intended ways). Particular techniques of police violence (use of firearms, use of batons, use of handcuffs) must be seen within these confines.

In this way, we can say 'police power' is a combination of mundane information techniques and mundane violence techniques (even where either or both move beyond formal regulation); 'police power' is not the result of a mysterious conspiracy. It is these techniques which make the existence and operations of modern police possible.

'The politics of the police', then, is a summary term for the many contests which have flourished and continue to flourish over these techniques. Should police forces exist? How should they exist? Should they be accountable to parliament? Should they produce individual reports of their actions? Should they use informers? How should they use informers? Should they use batons? How should they use batons? Should they use firearms? How should they use firearms? Should they use motor vehicles? How should they use motor vehicles? (Hogg 1987, 1988; Styles 1987). This list of questions at the core of the ongoing politics of the police could go on. What is important about any such list is that the issues in it signal a technical politics of the police which is part of the mechanism whereby the police contribute to the perpetual legal governance of societies, part of the 'attempt – incompleteness/failure – attempt' cycle.

Resistance to or by the police is also a technical matter for the sociology of law as governance. Resistance includes the mundane resistance involved in officers deliberately cutting corners in administrative procedures, people driving round the block to avoid a police breathalyser unit, officers claiming illegitimate expenses, or giving false names when questioned as well as the more spectacular resistance involved in people opposing police use of firearms, people protesting about police violence, officers using excessive violence against protesters, people shooting police in order to avoid arrest. Resistance to or by police also includes the 'imperative to resist'. This can occur in mundane resistance such as officers changing bits of their uniform just to be different, officers driving too fast just for the thrill of it, people abusing police just because they are authority figures, and in more spectacular resistance, people killing police for a thrill or police bashing prisoners for pleasure.

Policing as legal governance always involves knowledge in the selection of objects for policing – a contingent, historical knowledge – and the actual policing of those objects. This process is part of the administration of information we discussed above. The use of knowledge is a defining aspect of modern policing. Policing involves simple, definitional knowledge – humans organise things into property, modern organisations involve administration – which allows access to social objects. And policing involves more complex knowledge, used in attempting greater control or direction on these objects, for example fingerprints and informants' information may help apprehend a bank robber, reports to senior management need statistical evidence. Each technique of policing as legal governance, whether it is handcuffing, use of informants, report writing, interrogation of suspects, involves a combination of rational and irrational knowledge; thus, for example, knowledge such as a belief in the evil of 'villains' or an irrational fear of things going wrong, works alongside very rational procedural knowledge.

The knowledge involved in modern policing is always available knowledge – whether it is the dominant rational knowledge of computerised policing or the occasional turn to a clairvoyant to help solve a murder, the police can only use knowledge made available by existing institutions. The knowledge involved in modern policing is increasingly reliant on the human sciences and their attendant commitment to statistics. Modern policing uses psychology, sociology, biology, and of course criminology, alongside modern medical knowledge as it goes about its business. It features statistics in all its central work; indeed, a modern police force is almost unimaginable without crime statistics and other statistical measures of human behaviour.

Our third and final cluster of principles contains two method principles, concerned with distinguishing the subdiscipline's basic products from the uses of these products and its reflexivity. As these are little different for the study of policing than for the study of any other object, we need spend little time on them. Suffice to say that in studying the police, the sociology of law as governance must be especially careful and reflexive as the situation presents so many opportunities for the subdiscipline's products to be conflated with their uses. The details compiled in line with the careful generalisations about policing are obviously very useful to the police as they go about their business. This allows the strong possibility for practitioners of the subdiscipline to push improperly for new generalisations, or to try to exclude other practitioners as, for example, pressure is applied to a university department to provide special assistance to the police in exchange for greater funding.

In many respects the sociology of law as governance's approach to the study of the police is not dramatically different from other sociological approaches to this topic. As we indicated in the previous chapter, this is not a cause for concern; quite the reverse. In examining Foucault's work on law and in building a framework for a new sociology of law as governance out of this examination, we have been clear all along that the name Foucault does not signal an intellectual revolution. Rather, we have shown that it signals an opportunity to return to some nineteenth- and early twentieth-century sociological endeavours; hence our keenness to use Foucault's insights alongside those of Durkheim. The best way to use Foucault's work, we suggest, is as an instrument for ground clearing, surveying and mapping; such is our approach to Foucault and law.

Notes

Chapter 1: An Introduction to Foucault

1. The occasion on which Foucault comes closest to providing such a general synthesis is in 'Two Lectures' (TL 1980); these lectures are considered in more detail below.
2. The existence of 'many Foucaults' is not only evident in his writings but in the ever expanding secondary literature. In these works we are offered a choice between many 'Foucaults'; as structuralist and anti-structuralist, as compatible with Marxism and as radically opposed, as modernist and as postmodernist.
3. Our technique for referencing Foucault's texts is as follows. Where relevant we refer in square brackets to the date of original publication of Foucault's texts; dates in round brackets refer to the English translation that has been relied upon. For frequently cited works we have adopted simple abbreviations that are listed in the References list at the end of the book.
4. Among the many overviews of Foucault which provide a detailed discussion of the phases of his work the following should be considered: Dreyfus and Rabinow 1982; Merquior 1985; Sheridan 1980; Shumway 1989. Another approach to Foucault is provided by biographies that link the personal to the intellectual: Eribon 1991; Macey 1993; Miller 1993. Another approach is provided in collections of interviews with Foucault; Foucault 1977; P/K 1980; Foucault 1988a; Foucault 1989.
5. Compare Foucault's 'conditions of possibility' with the interesting but underexplored feature in Marx where he speaks of 'conditions of existence'. Marx's focus is to identify the preconditions (the 'without which') whereas Foucault is more concerned to stress the absence of any necessary connection between social elements which account for the specificity of the always unique 'event'.
6. Formulations of this type can be found in the following passages in Foucault: P/K: 99–100, 142, 188–9, 202–3. Further discussion of the extent to which Foucault addresses these issues is to be found in Jessop 1986 and Minson 1980.
7. Foucault also makes use of a related but rather more organic metaphor when he imagines social relations as a system of 'nets'

or 'networks' (HoS: 45). His most vivid and enduring imagery is that of the 'capillary' system of power (P/K: 96–7).

8. Foucault does occasionally make use of the term 'regulation' in more than a descriptive way. For example, he identifies regulation as a characteristic form of government of the 'bio-politics of the population' (HoS 1978: 149). Just once he speaks of an historical state which he calls a 'society of regulation', but he does not return to develop this theme (G 1979: 21).
9. Similarly, in an interview from the same period Foucault replies somewhat ambiguously to a question about liberalism and disciplinary society: 'I don't do this in order to say that Western civilization is a "disciplinary civilization" in all its aspects' (*Remarks on Marx* 1991: 167).
10. For a useful discussion of Foucault's views on disciplinary society see Smart 1983: 72–3.
11. It is not implied that the law is in any strong sense the cause of significant social change; it may be that law is simply a visible index of such changes.
12. For further discussion of the issues involved in Foucault's treatment of 'strategy' see Gordon 1980, Hunt 1992, Minson 1980, Smart 1983, Wickham 1983.
13. For fuller discussion of Foucault's relationship to Marxism see Cousins and Hussain 1984, Smart 1983, Poster 1984. For some of Foucault's own reflections see Foucault 1991.

Chapter 2: Law and Modernity

1. This quotation comes from the 'Preface' to *History of Sexuality* vol. II that appears in Rabinow's valuable collection (Rabinow 1984: 333–9). It does not appear in the standard English translation of *The Use of Pleasure* (Foucault 1985a).
2. The concept 'field of force relations' is one of Foucault's most elusive concepts. It is linked to his diagrammatic metaphor of power relations but he never explicates the connection between 'force' and 'power' (for discussion see Weedon 1987: 110–11).
3. Later Foucault presented an amended three-stage typology of forms of government: (1) 'the state of justice' of the Middle Ages corresponding to a 'society of laws'; (2) the 'administrative state' of the fifteenth and sixteenth centuries 'corresponding to a society of regulation'; (3) 'governmental state' corresponding to a type of society controlled by 'apparatuses of security' (G 1979: 21). The implications of this revised history of government are discussed below.
4. The term 'juridification' is sometimes used to suggest simply a quantitative expansion of law ('more law'). A more significant sense of the concept refers to any situation in which some regulatory

mechanism becomes transformed into distinctively legal form. For example, when the rules of some sports cease to be part of an oral tradition and get written down and include rules about subsequent amendments to the rules or where an employer issues a document laying down a grievance procedure for employees, these processes exemplify juridification. The idea of juridification was first used by Otto Kirchheimer to indicate the way in which law comes to be used as a means of neutralising political conflicts by subjecting them to formal legal regulation (Kirchheimer 1969). More recently the idea has come to refer to the process by which the state intervenes in areas of social life in ways which limit the autonomy of individuals or groups to determine their own affairs; see, for example, Habermas's discussion of 'Tendencies of Juridification' (Habermas 1987b: 356–73; Teubner 1986). These tendencies have led some commentators to worry about a new social disease of hyperlexis or legal overload (Trubek 1984: 824).

5. It is possible that Foucault's shift of attention towards governmental rationality, which occurred towards the end of the 1970s, involves the dropping or at least downgrading of his equation of pre-modernity with absolutism, law and sovereignty. Colin Gordon suggests that this was the case, but unfortunately he offers no textual evidence to support this view (Gordon 1987).
6. It is important to stress just how long the struggle for the franchise took; in broad terms the key period was that between 1830 and 1930, when full adult suffrage was secured, and even later in the United States (Therborn 1977).
7. The lectures have not, as yet, been translated into English, but are extensively summarised and discussed by Colin Gordon (Gordon 1991).

Chapter 3: Critique of Foucault's Expulsion of Law

1. It is interesting to note that Foucault comments on Kantorowicz's (1957) analysis of the king's 'two bodies' (the one being the physical body of the incumbent and the other the symbolic body of the Crown). Foucault manages to miss the point of this analysis which shows that it is only by virtue of these 'two bodies' that a conception of sovereignty can subsequently emerge which is *not* tied to the literal figure of a monarch.
2. While we make no claim to confirm the suggestion, there is a sense in which Foucault seems to have experienced a profound political disillusionment in the period of 'noımalisation' after 1968 (Eribon 1991: 274–7). The difficulty that confronted him and which he never made any direct attempt to resolve was that, politically, there was 'nowhere to go'. To swing to the right (as did André Glucksmann, one of his young associates during this period) was

for Foucault 'unthinkable'; it is perhaps not surprising that he espoused a certain rather loosely formulated anarchism, while at other times he avoided invitations to comment on the political implications of his intellectual positions.

3. For amplification of this point, along with many others about Foucault's treatment of punishment and prisons, see Garland 1990.
4. It should be noted that the juridification thesis comes in versions with very different political lineages; contrast Hayek's neo-liberalism (Hayek 1982) with Habermas's neo-socialist version (Habermas 1987b).

Chapter 4: Governance and its Principles

1. Our sketch of governmentality is heavily indebted to Gavin Kendall.
2. Our formulations regarding failure are heavily indebted to Jeff Malpas.

Chapter 6: Method Principles for the Sociology of Law as Governance

1. The formulation of this example owes a special debt to Albury (1989).
2. There are several versions of the derivation of this expression. As we understand it, it derives from the aftermath of a talk given early this century by the philosopher C.S. Peirce. After his talk Peirce was approached by a member of the audience. 'I enjoyed your talk,' she said, 'but it doesn't quite fit my theory of the universe.' 'Ah,' replied Peirce, 'and what might that theory be?' 'That the universe stands on the back of a giant elephant.' 'Very interesting,' the philosopher responded, 'but I must ask: on what does the elephant stand?' 'Easy,' his interlocutor answered calmly, 'it stands on the back of a giant turtle.' Fascinated, Peirce was about to continue his line of questioning, but didn't get the chance. 'Don't bother asking,' she politely suggested, 'it's turtles all the way down.'
3. See also Bauman (1989) for an interesting discussion of the ignoble ends to which the Nazis put some fairly standard, quite noble results of sociological work.
4. As Wittgenstein says 'it can never be our job to reduce anything to anything, or to explain anything' (1958: 18; see also Holmwood and Stewart 1991).

References

Writings of Michel Foucault

Where relevant the date of original publication is given in square brackets; dates in round brackets (with frequently cited works in abbreviation) refer to the English translation that has been relied upon.

Madness and Civilization: A History of Insanity in the Age of Reason [1964] (trans. A.M. Sheridan Smith) New York: Harper & Row, 1965 (M&C 1965).

The Order of Things: An Archaeology of the Human Sciences [1966] London: Tavistock, 1970 (OoT 1970).

The Archaeology of Knowledge and the Discourse of Language [1969] (trans. Alan Sheridan) London: Tavistock, 1972 (AoK 1972).

The Birth of the Clinic: An Archaeology of Medical Perception [1963] (trans. A.M. Sheridan) London: Tavistock, 1973 (BC 1973).

I, Pierre Rivière, Having Slaughtered My Mother, My Sister and My Brother: A Case of Parricide in the 19th Century New York: Pantheon, 1975.

Discipline and Punish: The Birth of the Prison [1975] London: Allen Lane and New York: Pantheon, 1977 (D&P 1977).

Language, Counter-Memory, Practice: Selected Essays and Interviews (ed. Donald Bouchard) Ithaca, NY: Cornell University Press, 1977.

The History of Sexuality vol. 1: *An Introduction* [1976] (trans. Robert Hurley) New York: Random House, 1978 (HoS 1978).

'Governmentality' 6 *Ideology & Consciousness* 5–21 (1979) and in Burchell *et al.*, 1991, pp. 87–104 (G 1979).

Power/Knowledge: Selected Interviews and Other Writings 1972–1977 (ed. Colin Gordon) Brighton: Harvester Press, 1980 (P/K 1980).

'Two Lectures' in *Power/Knowledge: Selected Interviews and Other Writings 1972–1977* (ed. Colin Gordon) Brighton: Harvester Press, 1980, pp. 78–108 (TL 1980).

'Omnes et Singulatim: Towards a Criticism of "Political Reason"' in S. McMurrin (ed.) *The Tanner Lectures on Human Values* (vol. 2) Cambridge: Cambridge University Press, 1981, pp. 223–54 (O&S 1981).

'The Order of Discourse' [1971] in R. Young (ed.) *Untying the Text: A Post-Structuralist Reader* London: Routledge, 1981, pp. 48–78 (OoD 1981).
'Is it Useless to Revolt?' 8 *Philosophy & Social Criticism* 1–9 (1981).
'The Subject and Power' in Herbert Dreyfus and Paul Rabinow *Michel Foucault: Beyond Structuralism and Hermeneutics* Chicago: University of Chicago Press, 1982, pp. 208–26 (S&P 1982).
The Foucault Reader (ed. Paul Rabinow) New York: Pantheon, 1984.
'Nietzsche, Genealogy and History' [1971] in Paul Rabinow (ed.) *The Foucault Reader* New York: Pantheon, 1984, pp. 76–9 (NGH 1984).
'On the Genealogy of Ethics: An Overview of Work in Progress' in Paul Rabinow (ed.) *The Foucault Reader* New York: Pantheon, 1984, pp. 340–72 (GE 1984).
The History of Sexuality vol. 2: *The Use of Pleasure* [1984] New York: Viking, 1985 (1985a).
The History of Sexuality vol. 3: *The Care of the Self* [1984] New York: Pantheon, 1985 (HoS 1985b).
'Technologies of the Self' in Martin Luther, Huck Gutman and Patrick Hutton (eds.) *Technologies of the Self* Amherst: University of Massachusetts Press, 1988a.
'Politics and Reason' in *Politics, Philosophy, Culture: Interviews and Other Writings 1977–1984* (ed. L.D. Kritzman) New York: Routledge, 1988 (1988b).
'The Dangerous Individual' in *Politics, Philosophy, Culture: Interviews and Other Writings 1977–1984* (ed. L.D. Kritzman) New York: Routledge, 1988 (1988c).
Foucault Live: Interviews 1966–1984 (ed. Sylvère Lotinger) New York: Semiotext(e), 1989.
Remarks on Marx: Conversations with Duccio Trombadori (trans. R. James Goldstein and James Cascaito) New York: Semiotext(e), 1991.

Other works

Abrams, Philip 1968 *Origins of British Sociology, 1834–1914* Chicago: University of Chicago Press.
Albury, Rebecca 1989 'Abortion? But I Thought That Was Settled Years Ago' 31/32 *Refractory Girl*.
Alexander, Jeffrey C. (ed.) 1988 *Durkheimian Sociology: Cultural Studies* Cambridge: Cambridge University Press.
Althusser, Louis 1969 'Contradiction and Overdetermination' in *For Marx* Harmondsworth: Penguin.
Austin, John 1955 *The Province of Jurisprudence Determined* [1832] London: Weidenfeld & Nicolson.
Bailey, Frederick G. 1983 *The Tactical Uses of Passion: An Essay on Power, Reason and Reality* Ithaca: Cornell University Press.

Bauman, Zygmunt 1989 *Modernity and the Holocaust* Cambridge: Polity Press.

Berman, Marshall 1982 *All That Is Solid Melts Into Air: The Experience of Modernity* New York: Simon & Schuster.

Bevis, Phil, Michèle Cohen and Gavin Kendall 1989 'Archaeologizing Genealogy: Michel Foucault and the Economy of Austerity' 18: 3 *Economy & Society* 323–45.

Burchell, Graham 1991 'Peculiar Interests: Civil Society and Governing "The System of Natural Liberty"' in Graham Burchell, Colin Gordon and Peter Miller (eds.) *The Foucault Effect: Studies in Governmentality* Hemel Hempstead: Harvester Wheatsheaf, pp. 119–50.

Burchell, Graham, Colin Gordon and Peter Miller (eds.) 1991 *The Foucault Effect: Studies in Governmentality* Hemel Hempstead: Harvester Wheatsheaf.

Carrithers, Michael, Steven Collins and Steven Lukes (eds.) 1985 *The Concept of the Person: Anthropology, Philosophy and History* Cambridge: Cambridge University Press.

Chapman, Brian 1971 *Police State* London: Macmillan.

Collins, Randall 1985 'The Durkheimian Tradition' in *Three Sociological Traditions* Oxford: Oxford University Press.

Cousins, Mark and Athar Hussain 1984 *Michel Foucault* New York: St Martins Press.

Curtis, Bruce 1992 *True Government By Choice Men?: Inspection, Education, and State Formation in Canada West* Toronto: University of Toronto Press.

Dandeker, Christopher 1991 *Surveillance, Power and Modernity* Cambridge: Polity Press.

David-Neel, Alexandra 1977 *Buddhism: Its Doctrines and Its Methods* [1939] New York: St Martin's Press.

Dawkins, Richard 1985 *The Blind Watchmaker* London: Longman.

Defert, Daniel 1991 '"Popular Life" and Insurance Technology' in Graham Burchell *et al.* (eds.) *The Foucault Effect*, pp. 211–34.

Demos, John 1982 *Entertaining Satan: Witchcraft and the Culture of Early New England* New York: Oxford University Press.

Donzelot, Jacques 1988 'The Promotion of the Social' 17: 3 *Economy & Society* 395–427.

Donzelot, Jacques 1991 'The Mobilization of Society' in Graham Burchell *et al.* (eds.) *The Foucault Effect*, pp. 169–80.

Dreyfus, Herbert and Paul Rabinow 1982 *Michel Foucault: Beyond Structuralism and Hermeneutics* Chicago: University of Chicago Press.

Duncombe, Jean and Dennis Marsden 1993 'Love and Intimacy: The Gender Division of Emotion and "Emotion Work"' 27: 2 *Sociology* 201–20.

Durkheim, Emile 1964 *The Rules of Sociological Method* [1895] New York: Free Press.

Durkheim, Emile 1965 *The Elementary Forms of Religious Life* [1912] New York: Free Press.

Dworkin, Ronald 1986 *Law's Empire* Cambridge, Mass: Harvard University Press.

Eribon, Didier 1991 *Michel Foucault* Cambridge, Mass: Harvard University Press.

Ewald, François 1990 'Norms, Discipline and the Law' 30 *Representations* 138–61, reprinted in Robert Post (ed.) *Law and the Order of Culture* Berkeley: University of California Press, 1991.

Ewald, François 1991 'Insurance and Risk' in Graham Burchell *et al.* (eds.) *The Foucault Effect*, pp. 197–210.

Fildes, Valerie 1988 *Wet Nursing: A History from Antiquity to the Present* Oxford: Blackwell.

Garland, David 1990 *Punishment and Modern Society: A Study in Social Theory* Oxford: Clarendon.

Gordon, Colin 1979 'Other Inquisitions' 6 *Ideology & Consciousness* 23–46.

Gordon, Colin 1980 'Afterword' to Colin Gordon (ed.) *Michel Foucault: Power/Knowledge* Brighton: Harvester, pp. 229–59.

Gordon, Colin 1987 'The Soul of the Citizen: Max Weber and Michel Foucault on Rationality and Government' in Scott Lash and Sam Whimster (eds.) *Max Weber, Rationality and Modernity* London: Allen & Unwin.

Gordon, Colin 1991 'Governmental Rationality' in Graham Burchell *et al.* (eds.) *The Foucault Effect*, pp. 1–51.

Habermas, Jürgen 1987a *The Philosophical Discourse of Modernity: Twelve Lectures* Cambridge, Mass: MIT Press.

Habermas, Jürgen 1987b *The Theory of Communicative Action:* vol. 2: *Lifeworld and System* Boston: Beacon Press.

Habermas, Jürgen 1993 *Facticity and Validity: Contributions to a Democratic Theory of Law and the Constitutional State* (trans. William Rehg) Cardozo Law Review, mimeo.

Hacking, Ian 1975 *The Emergence of Probability* Cambridge: Cambridge University Press.

Hacking, Ian 1990 *The Taming of Chance* Cambridge: Cambridge University Press.

Hacking, Ian 1991 'How Should We Do the History of Statistics?' in Graham Burchell *et al.* (eds.) *The Foucault Effect*, pp. 181–96.

Hall, Stuart 1988 'The Toad in the Garden: Thatcherism Among the Theorists' in C. Nelson and L. Grossberg (eds.) *Marxism and the Interpretation of Culture* Urbana: University of Illinois Press, pp. 35–73.

Hart, H.L.A. 1961 *The Concept of Law* Oxford: Clarendon Press.

Hay, Douglas 1975 'Property, Authority and the Criminal Law' in Doug Hay *et al. Albion's Fatal Tree: Crime and Society in Eighteenth Century England* London: Allen Lane.

Hayek, Friedrich A. 1982 *Law, Legislation and Liberty* London: Routledge.

Hirst, Paul 1986 *Law, Socialism and Democracy* London: Allen & Unwin.

Hogg, Russell 1987 'The Politics of Criminal Investigation' in Gary Wickham (ed.) *Social Theory and Legal Politics* Sydney: Local Consumption, pp. 120–40.

Hogg, Russell 1988 'Police "Hot" Pursuits: The Need for Restraint' in Michael Hogan *et al.* (eds.) *Death in the Hands of the State* Sydney: Redfern Legal Centre Press.

Holmwood, John and Alexander Stewart 1991 *Explanation and Social Theory* London: Macmillan.

Humphreys, Christmas 1949 *Zen Buddhism* London: Heinemann.

Hunt, Alan 1978 *The Sociological Movement in Law* London: Macmillan.

Hunt, Alan 1992 'Foucault's Expulsion of Law: Towards a Retrieval' 17: 1 *Law & Social Inquiry* 1–38.

Hunt, Alan 1994 *Governance of the Consuming Passions: A History of Sumptuary Regulation* London: Macmillan, forthcoming.

Jackson, Stevi 1993 'Even Sociologists Fall in Love: An Exploration in the Sociology of Emotions' 27: 2 *Sociology* 221–42.

Jessop, Bob 1986 'Poulantzas and Foucault on Power and Strategy' 3 *Ideas & Production* 59–84 and in Bob Jessop 1990 *State Theory: Putting the Capitalist State in its Place* Cambridge: Polity Press.

Kantorowicz, Ernst 1957 *The King's Two Bodies: A Study in Medieval Political Theology* Princeton: Princeton University Press.

Katz, Jack 1988 *Seductions of Crime: Moral and Sensual Attractions in Doing Evil* New York: Basic Books.

Kirchheimer, Otto 1969 *Politics, Law and Social Change: Selected Essays of Otto Kirchheimer* (eds. F.S. Burin and K.L. Shell) New York: Columbia University Press.

Langbein, John 1977 *Torture and the Law of Proof: Europe and England in the Ancien Regime* Chicago: University of Chicago Press.

Luhmann, Niklas 1985 *A Sociological Theory of Law* London: Routledge & Kegan Paul.

Macaulay, Stewart 1963 'Non-Contractual Relations in Business: A Preliminary Study' 28 *American Sociological Review* 55–67.

Macey, David 1993 *Michel Foucault* London: Hutchinson.

Marshall, T.H. 1963 'Citizenship and Social Class' in *Sociology at the Crossroads* London: Heinemann, pp. 67–127.

Mauss, Marcel 1985 'A Category of the Human Mind: The Notion of Person; the Notion of Self' in Michael Carrithers, Steven Collins and Steven Lukes (eds.) *The Category of the Person: Anthropology, Philosophy, History* Cambridge: Cambridge University Press, pp. 1–25.

Merquior, Jose 1985 *Foucault* London: Fontana Press.

Miller, James 1993 *The Passion of Michel Foucault* New York: HarperCollins.
Miller, Peter and Nikolas Rose 1990 'Governing Economic Life' 19: 1 *Economy & Society* 1–31.
Minson, Jeffrey 1980 'Strategies for Socialists? Foucault's Conception of Power' 9 *Economy & Society* 1–43.
Minson, Jeffrey 1985 *Genealogies of Morals: Nietzsche, Foucault, Donzelot and the Eccentricity of Ethics* London: Macmillan.
Nisbet, Robert 1965 *Emile Durkheim* Englewood Cliffs: Prentice-Hall.
O'Malley, Pat 1991 'Legal Networks and Domestic Security' in Austin Sarat and Susan Silbey (eds.) *Studies in Law, Politics and Society* vol. 2 Greenwood, Conn: JAI Press.
O'Malley, Pat 1992 'Risk, Power and Crime Prevention' 21:3 *Economy and Society* 252–75.
O'Malley, Pat 1993 'Containing Our Excitement: Commodity Culture and the Crisis of Discipline' in Austin Sarat and Susan Silbey (eds.) *Studies in Law, Politics, and Society* vol. 13 Greenwood, Conn: JAI Press, pp. 159–86.
Palmer, Jerry and Frank Pearce 1983 'Legal Discourse and State Power: Foucault and the Juridical Relation' 11 *International Journal of the Sociology of Law* 361–83.
Pasquino, Pasquale 1991 'Theatrum Politicum: The Genealogy of Capital – Police and the State of Prosperity' in Graham Burchell *et al.* (eds.) *The Foucault Effect*, pp. 105–18.
Poster, Mark 1984 *Foucault, Marxism and History: Mode of Production versus Mode of Information* Cambridge: Polity Press.
Poulantzas, Nicos 1978 *State, Power, Socialism* London: New Left Books.
Rabinow, Paul (ed.) 1984 *The Foucault Reader* New York: Pantheon.
Rinpoche, Sogyal 1992 *The Tibetan Book of Living and Dying* London: Rider Books.
Rose, Nikolas 1989 *Governing the Soul: The Shaping of the Private Self* London: Routledge.
Rose, Nikolas 1991 'Governing by Numbers: Figuring Out Democracy' 16: 7 *Accounting, Organization and Society* 673–92.
Rose, Nikolas 1992 'Governing the Enterprising Self' in P. Heelas and P. Morris (eds.) *The Values of the Enterprise Culture: The Moral Debate* London: Routledge.
Rose, Nikolas and Peter Miller 1992 'Political Power Beyond the State: Problematics of Government' 43: 2 *British Journal of Sociology* 173–205.
Rosen, George 1974 *From Medical Police to Social Medicine: Essays on the History of Health Care* New York: Science History Publications.
Santos, Boaventura de Sousa 1985 'On Modes of Production of Law and Social Power' 13 *International Journal of the Sociology of Law* 299–336.
Sheridan, Alan 1980 *Michel Foucault: The Will to Truth* London: Tavistock.

Shumway, Michel 1989 *Michel Foucault* Boston: Twayne Publishers.
Simon, Jonathan 1987 'The Emergence of a Risk Society: Insurance, Law, and the State' 95 *Socialist Review* 61–89.
Smart, Barry 1983 *Foucault, Marxism and Critique* London: Routledge & Kegan Paul.
Smith, Richard S. 1985 'Rule-by-records and Rule-by-reports: Complementary Aspects of the British Imperial Rule of Law' 19: 1 *Contributions to Indian Sociology* 153–76.
Somers, Margaret 1993 'Citizenship and the Place of the Public Sphere: Law, Community, and Political Culture in the Transition to Democracy' 58 *American Sociological Review* 587–620.
Spierenburg, Pieter 1984 *The Spectacle of Suffering: Executions and the Evolution of Repression* Cambridge: Cambridge University Press.
Styles, John 1987 'The Emergence of the Police: Explaining Police Reform in Eighteenth and Nineteenth Century England' 27: 1 *British Journal of Criminology* 15–22.
Teubner, Gunther (ed.) 1986 *Dilemmas of Law in the Welfare State* Berlin: Walter de Gruyter.
Therborn, Göran 1977 'The Rule of Capital and the Rise of Democracy' 103 *New Left Review* 3–41.
Trubek, David 1984 'Turning Away From Law' 82 *Michigan Law Review* 824–35.
Weber, Max 1954 *Law in Economy and Society* (ed. Max Rheinstein) Cambridge, Mass: Harvard University Press.
Weedon, Chris 1987 *Feminist Practice and Poststructuralist Theory* Oxford: Blackwell.
Whitton, Joel and Joe Fisher 1986 *Life Between Life* New York: Warner Books.
Wickham, Gary 1983 'Power and Power Analysis: Beyond Foucault?' 12: 4 *Economy & Society* 468–98.
Wickham, Gary 1987 'Turning the Law into Laws for Political Analysis' in Gary Wickham (ed.) *Social Theory and Legal Politics* Sydney: Local Consumption Publications.
Wittgenstein, Ludwig 1958 'The Blue Book' in *The Blue and the Brown Books* Oxford: Blackwell.